# TRANSFORM

## or

# DIE

The Ultimate CEO Playbook for AI Collaboration

# TRANSFORM
## or
# DIE

## How to Build Teams that Outthink, Outpace, and Outprofit the Competition in the AI Age

Complimentary Access to the K.E.R.N.
Human+AI Collaboration® Online Toolkit

## Russell M. Kern

Author Name Russell M. Kern
Author Email Address russell@kernandpartners.com
Author's website https://www.kernandpartners.com/

Transform or Die, Russell M. Kern—1st ed.
ISBN  Paperback: 979-8-89764-130-7
       Hardback: 979-8-89764-131-4

CERTIFIED

(H)

WRITTEN
BY HUMAN

# Testimonials

Russell has been an extraordinary business leader, entrepreneur, and friend. He built a thriving agency that he later sold to Omnicom, where I had the privilege of working alongside him for several years. His passion for leadership consistently inspires those around him and brings out the very best in every colleague and partner.

Robert G. Horvath
President
D'Addario

Artificial Intelligence is not the end of something, but the widening of the canvas. As Transform or Die makes clear, the organizations that thrive will be those that treat AI as a partner in discovery—an ally in expanding human potential, not diminishing it. Russell Kern reminds us that reflection, more than reaction, is the leader's most powerful instrument. Transformation is less about defense against change, and more about the courage to embrace it fully.

Gerald A. Bagg
Co-Founder
Quigley-Simpson

Continuous business growth always requires people who are willing to embrace that change is the only constant in our industry (and life). Yet that change for the good of business and its people can only be accelerated if leadership is willing to transform the business regardless of their past successes.

Russell with over 40 years as an entrepreneurial CEO has never stopped embracing change with a willingness to adapt his perspective on what is right for today and tomorrow thereby not by relying on the past.

A real gift which very few experienced CEO's have or willing to embrace. As such he is perfectly placed to embrace the value of human + AI to create sustainable growth.'

Marco Scognamiglio
Start Up Growth Advisor for People-Based Businesses
Former Global CEO of RAPP Worldwide

Transform or Die is a great example of applying Appreciative Inquiry methodology (the "other AI') to the increasing challenges of AI adoption and team collaboration. Russell's approach embodies the principle that 'the art of asking unconditionally positive questions to surface strengths-in-practice' can drive organizational transformation. His 'Empower' pillar particularly reflects our research on building generative capacity.

Russell shows how leaders can create conditions where teams naturally and continuously innovate and pursue excellence together by focusing on what's possible, not what's broken, with a collaboration platform that embraces AI.

Ronald Fry, PhD
B Charles Ames Professor of Management
Case Western Reserve University
Cleveland, Ohio

True business growth comes when marketing teams work in lockstep—aligning strategy, creativity, and execution across every channel. When collaboration is strong, we're able to accelerate brand impact, shorten go-to-market cycles, and drive measurable ROI. Throughout his storied career, Russell has a track record and proven frameworks to break silos, optimize processes, and unlock the full potential of any team's collective talent.

Craig Lister
Chief Marketing Officer
National Wireless Carrier

Russell is one of those rare leaders who has a CEO side hustle of rapidly democratizing the new knowledge he has gained to accelerate his colleague's success. He provides practical strategy to execution frameworks that are timeless in their application through decades of industry change and technology innovation cycles.

Michele B. Weber
CMO/VP of Marketing
B2B Tech Industry

The AI age is upon us, and Russell Kern's new book, "Transform or Die" is a timely and important guide that should be on every leader's bookshelf. It is a vital tool if leaders expect to drive the success of their teams in this new environment. Russell's vast experience, business insights, and passion for people and teams are the essential foundation for this provocative and practical book.

Thomas M. Doolittle, Ph.D.
Human Resources and Executive Coach
Former CAT Global HR Leader

Russell is one of those guys people want to work for - not for the money but because of who he is, and how he is. If anyone knows how to build, and get the best from, from teams, it's him. I know this because I've had the pleasure of working for him, and look forward to the next opportunity to do so. I've seen him attract talented people and turn them into a happy, highly effective team. There is much to learn from his experience and approach.

Sean Riches
Director
Erskine, Nash Associates , Ltd. UK

Leaders are facing the most significant shift in their business and careers because of AI. At the same time, they are also presented with the most significant opportunity. Russell Kern provides strategies and frameworks that help leaders prepare their teams and organizations to outthink, outpace, and outprofit their competition based on decades of leadership experience."

Andreas Welsch
Founder & Chief AI Strategist
Intelligence Briefing

The most important upgrade you can make in the way that others see you as a leader is to change the way you see yourself as a leader. Russell Kern knows his stuff, but more importantly, he knows your stuff. He knows what's holding you back and how to move you forward. Of course, Russell has that history of building great leaders and building great companies, but the thing that he has that makes him truly special is the heart and desire to pass that knowledge onto you.

Rob Goldenberg
Chief Creative Officer
Jewish Federation Los Angeles

Russell and I have known each other for years. He has consistently been an incredible sounding board of ideas. His years of managing businesses across a variety of teams yields tremendous insight and experience. These have proven valuable to me as I tackle a plethora of challenges across diverse areas. The common connection Russell brings together in his work and in his book is people; how to connect, inspire and help each person to thrive.

Matt Frankel
Director of Assay Research
Global Medical Diagnostic and Pharma Manufacture

In an era where AI is reshaping every industry, this book cuts through the noise with uncomfortable truths about why most organizations are failing to adapt. Russell doesn't just identify the myths that keep teams stuck in outdated thinking—the K.E.R.N. Framework provides a practical roadmap for building the kind of human-AI collaborations that will separate thriving organizations from those left behind. Essential reading for any leader serious about their organization's survival in the AI age."

Joretha G. Johnson
CEO
Advanced Transformational Technologies

Having worked with and coached dozens of Mid-Market CEOs as a Vistage Chair, Russell is the go-to expert on collaboration development for high-performance teams. This book, Transform or Die, captures the step-by-step methodology he's refined through decades of real-world experience and building a $100MM company. His Human+AI framework will help leaders avoid collaboration development blind spots while reducing the risk from costly AI pilot failures. The key point Russell makes, which I fully support, is that business growth, profits, and success are not about the tech; it's always about leadership, trust, and effective collaboration.

Sandeep Khera
CEO Peer Group Chair & Executive Coach
Vistage and TIGER 21

Nature shows us that anything not growing is dying or already dead. The same principle applies to you and me. We MUST grow! It is the nature of life.

Russell Kern has grown many businesses, and he can expertly guide yours too. I know and admire him as a business leader and a veteran corporate executive. I also consider him a friend. Read this book in pieces. Apply what you've learned, then read some more. You'll grow!"

Jim Cathcart, CSP, CPAE
Mentor to Professional Experts
CEO of Cathcart.com

I feel like an immigrant to the 21st century. So much has changed and continues to change faster each year. How do we stay relevant? I am inspired by your enthusiasm for artificial intelligence and your ability to synthesize and explain its complexities.

Charles Darwin says that the species with the highest probability of survival is the one that adapts the fastest - not the smartest or the strongest. You show us the opportunity in the uncertainty by embracing radical technology, surrounding yourself with young professionals, keeping your sense of purpose and by staying curious.

There are many jewels in the mouth of the dragon. Your book is an instruction manual and a way forward adapting to change. When I was just starting out you handed me a signed copy of "The Richest Man in Babylon." "Transform or Die" will be right beside it in my bookshelf. Both highlight your strength as a natural teacher, leader, and mentor. Bottom line. Go fast. Take chances.

Rob Pasnau
Managing Director
DMG Exports, INC. (Raleigh Bicycles USA
and Latin America)

Having seen Russell grow his business from a start up to a thriving multi-million dollar company, the secret sauce that Russell always brings to leaders is unabated honesty about what he sees, good or bad, in their business. Whether it relates to his own services or not, leaders from the many Fortune 500 clients he works with have trusted Russell's counsel for years. That kind of ego-free, drama-free, agenda-free advice is why his clients stay with him no matter how big or small their business or their challenge.

Camilla Grozian-Lorentzen
Founder & CEO
Lorentzen Marketing

I have known Russell for over 20 years and had the privilege of working closely with him for many of those years. He is a true entrepreneur—always seeking innovative ways to drive efficiency, foster collaboration, and unlock value for teams of all sizes. His forward-thinking approach and deep expertise make him an invaluable resource for any leader looking for new ways to achieve success.

David Azulay
Global Business Lead
World's Largest Ad Agency Holding Company

I've known Russell for decades, and he's consistently been the most focused, driven, and insightful person in any room. His depth of understanding, especially in emerging technologies, is unmatched. If he's coaching CEOs on AI strategy, then I am confident his book is a must-read.

Brian Johnson
Partner, Vanguard IP

Russell has the rare ability to translate decades of leadership into practical, actionable insight, which has changed how I think as a leader. As he has done his entire career, this book combines vision and practical frameworks to prepare teams for the future without losing focus on what really drives results, people.

CJ Forse
Founder & CEO
Vulgar

# Dedication

To my wife, Meryl, your intelligence and unwavering support has always kept me pushing forward.

I am grateful for your encouragement throughout the lows and highs of business, life and this book's development process.

As in Charlie Maskey's Book, when the Mole asked the Boy, "What do you want to be when you grow up?"

My hope for our life together and this book is to spread the Boy's powerful response:

"Kind".

# Acknowledgements

Thank you to my daughter, Hillary, and her husband, Lee, for their valuable input and guidance on the early chapters.

I also want to acknowledge the other children in my life, Matt, Maya, Lindsey, Steve, Sean and Ericka.

A special thanks goes to my brother, Dr. Ken Kern, for our many strategic brainstorming deli breakfast sessions.

Thank you to my brother, Dr. Morton Kern, and my sister Dr. Gail Acuna for their humor and love. In memory of my parents, Dr. Arthur and Bea Kern, for their gift of encouraging lifelong learning.

To my many friends (Bill, Tom, Brian, Paul, Allen, Frank, Lenny) who have stuck with me throughout the decades of my life, the ideation and the development process of the K.E.R.N. Human+AI Collaboration Framework®, I appreciate your care.

To my incredible clients within AT&T, AMEX, AAA, Blue Cross, Canon, DirecTVMerck, Sprint, SAP, T-Mobile and many others, thank you for the gift of our working relationship over the decades.

I want to thank (Paul, Brad, Craig S, Craig L, Jess, Julie, Mark, Suzanne, Stuart, Tina) every client I've had the privilege of working with. You are the best and brightest strategic, marketing, creative, operations, sales, data & analytics, and human behavior minds in the world.

Thank you to all my former Kern Agency team members (Camilla, David A, Zeke, Lynne, Desmond, David V., Nick, and hundreds more), support partners, as well as the Omnicom Organization. You helped the KERN Agency grow to levels we never dreamed of. And a special note of appreciation to Patrica Bare, for her unwavering tenacity and detail-minded support over the decades.

I want to acknowledge my mentors, coaches and colleagues, Bob Hacker, Zeke Ibarbia, Bob Horvath and Matt Halfkin, Marco Scognamiglio, Leigh Ober, for their brilliance and guidance. The sharing of their time and wisdom allowed me to find many pathways of success over my life's journey.

Finally, I want to thank Rod and Kathryn Jameson (may their memory be a blessing), who were Founders of my Summer Camp, Jamesons Ranch Camp, outside of Bakersfield, CA.

I am especially grateful for their life-guiding words, shared with every new camper at the start of every camp session: "You get out of camp what you put into it." It taught me this is the same philosophy about life.

# CONTENTS

# Preface

---

*"The intelligence to power our brain requires just 20 watts of power, equivalent to a dim light bulb, and operatesunderwater. It is estimated that the energy required to mimic the intellectual capabilities humans possess is equal to the energy a million homes consume."*
~ Inspired by the article, "The Future of AI Lies in Monkeys, Not Microchips" by Cory Miller[1]

---

## Opening Comments

Throughout every era of human progress, there have been moments when a new, unfamiliar technology emerged, initially unsettling people but ultimately transforming how we live and work.

Artificial Intelligence is rapidly emerging as one of these forces, poised not only to reshape entire industries but to radically expand what is possible for human creativity, innovation, and fulfillment at work.

---

[1] https://www.wsj.com/opinion/
the-future-of-ai-lies-in-monkeys-not-microchips-c855aad6

Predictions suggest that the widespread adoption of AI could boost global GDP by as much as $15 trillion in the coming decade[2], rivaling historic leaps such as the advent of electricity, the automobile, or powered flight. For every job or profession disrupted by these breakthroughs, history also shows us that countless new roles and opportunities have arisen that we once thought unimaginable.

It is only natural to feel trepidation in the face of such change; as humans, we are wired to crave predictability and certainty. Yet, embracing the unknown has always been the catalyst for humanity's greatest advancements. AI, if harnessed thoughtfully and ethically, holds the promise not just of efficiency, but of amplifying our uniquely human strengths—imagination, empathy, and purposeful problem-solving—far beyond what we could achieve alone.

I wrote this book from the perspective of a CEO who has spent forty years building and leading a nationally recognized advertising agency, navigating the complexities of growth, talent development, and lifelong learning. My direct experience creating the KERN College and pursuing advanced certifications in learning and development has deeply informed my approach, not only to adapting to change, but to helping others thrive through it.

With this book, I invite you to envision and actively shape a future where the process of humans working in collaboration

---

2    Gerrard, Juliet A, et al. "By 2030, AI Will Contribute $15 Trillion to the
     Global Economy." World Economic Forum, 7 Aug. 2019, www.weforum.org/
     stories/2019/08/by-2030-ai-will-contribute-15-trillion-to-the-global-economy/.

with AI does not replace workforces but multiplies them. Let us step forward together, embracing the possibilities that lie ahead.

**The mission of this book is threefold:**

- To provide a clear framework for improving communication and collaboration within and across teams

- To guide leaders in embracing the transformative potential of the range of AI technologies in the workplace

- To offer specific, step-by-step strategies for transforming how your organization partners with AI as a new, powerful team member who helps everyone thrive and flourish

Throughout, I walk my talk by fully embracing AI as my strategic partner in ideation and writing throughout the creative journey of this book, from the initial outline and tool creation, all the way to publication.

With many moments of triumph along the way and a new depth of knowledge gained, I used myself as a test case. How hard is AI to learn? What are the most common mistakes a CEO will make when using AI? What's the difference between them? Can AI really be a strategic collaboration partner? By actively collaborating with AI tools on a daily basis, I engaged in an iterative process—moving from human input to AI-generated output, back to human refinement, and then returning to AI suggestions for further enhancement.

I leveraged AI for brainstorming, research, drafting initial versions, evaluating and refining my writing, positioning the book, conducting keyword analysis, confirming solutions for my target audiences, and even inspiring the development of the four pillar K.E.R.N. Human+AI Collaboration Framework.

I tested dozens of open-source AI chatbots to compare their outputs, engineered literally hundreds of prompts, and reviewed thousands of pages of results. Through this process, I carefully incorporated content that I found to be accurate, significant, and engaging to read. Along the way, I was amazed and delighted to find many of the AI-generated outputs provided extraordinary inspiration, sparking ideas I would never have conceived on my own. Conversely, I also found myself wasting hours not fully understanding that AI is a machine designed to "please" and thus is prone to hallucination, making up answers that feel real but are entirely false.

As you are about to embark on your own discovery and development journey, I'd like to share a few insights I learned to help you gain the most you can from the pages ahead.

## Lesson 1: The Term "Artificial Intelligence" Is a Misnomer

While the technology is "artificial" in that it seeks to mimic human abilities, its outputs are far from truly "intelligent."

AI-generated responses are determined entirely by the data and training of the specific model. The software behind the user interface is designed to provide the best possible response to a given prompt by calculating the most probable sequence of words, image pixels, or sound waves based on its training data. Essentially, AI functions as a sophisticated prediction engine, ordering outputs according to patterns it has learned.

Therefore, the "intelligence" implied in the term "artificial intelligence" is misleading. AI does not possess human-level intelligence, nor does it have the five senses that allow humans to intake and process information at extraordinary speeds through our complex nervous systems.

However, AI remains a powerful tool for knowledge. It can rapidly find, retrieve, and analyze vast amounts of information, enabling humans to discover new words, ideas, suggestions, and patterns with remarkable speed.

## Lesson 2: The Journey with AI is Just Beginning

The pace of technological advancement in AI is unprecedented, reshaping industries and creating opportunities that few could have imagined even several years ago. While large language models have been developing quietly behind the scenes for nearly a decade, their recent leap into the mainstream—thanks to innovative chatbots and massive investments in AI infrastructure—means powerful new tools are now appearing in our daily lives at an astonishing rate.

As Marc Benioff (Founder and CEO of Salesforce.com) observed, "We are talking about a $12 trillion digital labor opportunity from agentic AI that will impact every business in every geography in the world." The sheer scale and impact of these changes can seem overwhelming, but they also open doors for individuals and organizations willing to embrace continued learning.

Testing new tools, exploring creative applications, and even making mistakes will deepen your understanding far more effectively than simply observing passively. By viewing your relationship with AI as a journey rather than a destination, you unlock endless opportunities for innovation and growth for yourself, your team, and your entire organization.

## Lesson 3: Why Human Oversight Is Irreplaceable When Using AI

Even as AI tools grow more powerful, human judgment remains indispensable whenever you rely on AI-generated content. While AI tools can draft text at remarkable speed, they are also known to produce hallucinations—answers that look plausible on the surface but are actually incorrect or even fabricated, sometimes inventing convincing details and references out of thin air. Relying solely on AI outputs, without a careful review, exposes your work to significant risks, including costly factual errors and a loss of credibility.

When you engage with AI tools such as Claude, ChatGPT, Gemini, Perpexity, Grox, and Copilot, it's easy to be impressed by how quickly they provide information or help you generate new material. However, efficiency does not equal accuracy. For anyone using AI to assist with research, writing, or decision-making, it is crucial to treat each AI-generated response as a first draft rather than a finished product.

To get the most from AI, see it as a collaborative assistant, not a replacement for critical thinking. Always validate facts, double-check sources, and use your experience to filter out dubious outputs. By pairing the speed of AI with human discretion, you amplify both creativity and reliability, resulting in better outcomes for your readers, team, and organization.

## Lesson 4: Prioritize Source Data and Data Security

**Data security, intellectual property protection, and cybersecurity should be at the core of every AI strategy, especially when it comes to open-source tools like ChatGPT, Claude, Gemini, Perplexity, and Deepseek.** Using these technologies without robust safeguards can carry significant risks to your organization's proprietary information, customer data, and financial details.

Before introducing or expanding the use of open-source AI tools on your team, work closely with your technical leadership (CTO, CIO, or Chief AI Officer) to ensure there are clear boundaries around sensitive data.

Be sure to investigate whether the data you are using is the right data, given the objective and mission of using AI.

Unfortunately, even if your organization hasn't formally granted access to open-source tools, SHRM reports[3] that up to 78% of the workforce uses various AI tools. This Shadow AI use means that your workforce may be uploading information that should remain confidential. Most of the time, this happens out of curiosity rather than ill intent, as people seek ways to enhance productivity.

**To keep everyone—and your company—safe:** Implement clear, practical AI-use protocols. Develop guidelines that are straightforward for every employee to understand and apply, regardless of their busy schedule.

Adopt a straightforward data classification system. Make it easy for staff to know:

- What data must always stay within secure company systems (e.g., any sensitive or confidential information)
- What data may be safely shared or transferred externally
- What data falls in between and needs case-by-case evaluation

Create an actionable checklist that fits your workflow. A concise three-category list empowers employees to quickly recognize which data can be used with AI tools and which data should be kept private.

---

3   Zielinski, Dave. "Shadow AI Is on the Rise: Why It Matters to HR." Shrm.org, 2025, www.shrm.org/topics-tools/flagships/ai-hi/shadow-ai-on-the-rise. Accessed 12 Aug. 2025.

Build a culture that rewards transparency. Encourage team members to speak up if they're unsure or if a mistake occurs. Early detection is the most effective way to prevent minor missteps from escalating into major breaches.

Finally, work in partnership with technical leaders to craft backup plans and ensure operational continuity. No system is perfect, and there will be times when human judgment falls short of expectations. With the proper preparation, you can confidently foster innovation while keeping your data—and your organization—secure.

## Lesson 5: Understanding AI Technology Matters More Than Most Leaders Realize

While every leader faces constant time pressures and an overwhelming influx of information, it can feel nearly impossible to carve out time to learn a new technology. However, I have discovered that the more I understand about AI, the less I know. The more you understand what AI models are, how they are built, programmed, and function, the easier it becomes to see how they can be used strategically in your business. On the other hand, you realize there are levels of complexity that require experts to be included in your strategic thinking about your business and what processes or workflows can benefit from being reengineered through an AI lens.

Just as with acquiring any new skill or learning a foreign language, I encourage leaders to develop at least a basic understanding of the data sets used to train their AI models. They need to be aware of the inherent biases within those models, as well as how the data flows into and out of their systems over time.

I am not suggesting that leaders must become experts in AI technology. However, the more foundational concepts you grasp, the more effective—and more meaningful—your conversations will be with your technology teams as you expand AI adoption across your organization.

## How This Book Is Organized

### Section 1: Context and Rationale (Chapters 2–6)

This section provides essential context for the many facets of collaboration. We will explore why true team success demands more than simply placing the right people in the right roles and examine the neuroscience behind effective teamwork. You'll learn how to leverage the brain's natural processes to enhance collaboration. In Chapter 4, we will discuss how to assess your team's current collaboration levels, the hidden costs of low collaboration, and the tangible benefits of building high-collaboration environments.

### Section 2: The K.E.R.N. Framework (Chapters 7–11)

Here, you'll be introduced to the K.E.R.N. Human+AI Collaboration Framework, followed by an in-depth exploration of its four pillar principles: Know, Empower, Reflect, and Nurture. This roadmap includes perspectives on approaching each principle from both human and technological standpoints, offering practical guidance for developing more powerful, future-ready teams.

**Section 3: Getting Started (Chapters 12 and Conclusion)**
This final section examines the common challenges faced by senior leaders today and how AI can serve as a strategic partner in overcoming them. You'll find resources for evaluating potential AI vendors, real-world examples of successful business use cases, and a range of ideas to jump-start internal ideation and collaboration on AI experiments and pilots. Additionally, you'll receive practical Human+AI training outlines you can adapt for your own organization.

**Given the incredible adoption and speed at which AI is coming to market, this book will quickly find itself dated.**

Given the rapid changes and evolution of AI Technologies, as a thank you for your time and readership, I am providing all readers access to Transform or Die as a living, consistently updated reference resource where you will find the latest information about AI tools and platforms, practice prompts, easy-to-use scorecards, and self-directing training programs for your teams.

Use this QR code or visit RussellMKern.com/transform to access the resources.

# Chapter 1

# Why Traditional Team Building Falls Short in the Era of Human+AI Collaboration

---

*"Every business is at an inflection point in business like we've never seen before."*

~ Marc Benioff - Founder, CEO of Salesforce

---

History offers stark reminders of how inadequate communication and collaboration can devastate even the world's most esteemed organizations. Tragedies like the Space Shuttle Challenger explosion and the Boeing 737 MAX disasters serve as powerful warnings for today's leaders.

It's easy to think, "That's not my organization. Those are rare exceptions." Yet the data tells a different story: 75% of cross-functional teams struggle with dysfunction[4], and just 7% of U.S. employees strongly agree that workplace communication is effective. The consequences are steep—David Grossman reported in 'The Cost of Poor Communications' that a survey of 400 companies with 100,000 employees each cited an average loss per company of $62.4 million per year because of inadequate communication to and between employees.[5]

## Why Traditional Team Building Falls Short in Today's Workplace

The accelerating pace of change, increasing uncertainty, shifting generational expectations, and widening knowledge gaps left by retiring workers all demand a new approach to building effective team dynamics. Cultivating a work culture that prioritizes strong relationships and creates a psychologically safe environment for open dialogue is no longer optional—it's essential for innovation, resilience, and sustainable success.

Classic team-building activities—such as rope courses, off-site dinners, or cooking classes—may build camaraderie. However, they often fail to address the daily realities and complex challenges of modern work. Today's teams must contend with continuous change, hybrid and remote work structures, and

---

4   Tabrizi, B., Lam, E., Girard, K., & Irvin, V. (2017). The Secret to Leading Organizational Change Is Empathy. Harvard Business Review.

5   SHRM. "The Cost of Poor Communication." Www.shrm.org, 2016, www.shrm.org/topics-tools/news/organizational-employee-development/cost-poor-communication.

the growing influence of AI-driven digital labor. These issues cannot be solved with a simple trust fall.

Driving real behavior change in the workplace requires hands-on practice applied in near real-time to actual work tasks. Research suggests that 70% of adult learning comes from applying new knowledge directly on the job. Moreover, mastering the skills needed to collaborate effectively with various AI tools—as strategic partners that enhance human capabilities—demands both AI literacy and opportunities for experimentation across the organization.

## The Critical Issues That Keep Leaders Awake at Night

How many of these challenges do you recognize in your organization—or worry might already be undermining it?

- Teams and departments operating in silos
- Incentives that drive misalignment
- Leaders who fail to model collaborative behaviors
- Ineffective communication tools
- Mounting stress from constant time pressure and task overload
- Lack of trust between team members and a low feeling of safety to share opinions
- Resistance to change

The complexities of hybrid and remote work can further exacerbate these problems. Together, they risk derailing opportunities, stalling innovation, and disengaging employees—compounding the already serious threats to your organization's success.

## Why Is It So Difficult to Drive Change?

Neuroscience shows that leaders and teams are unlikely to embrace change until the pain of maintaining the status quo becomes greater than the discomfort of transformation. For senior leaders, the challenge is compounded as organizational growth and accumulated successes provide continuous evidence reinforcing the belief that the current operating culture is effective. This psychological and neurological feedback loop creates a powerful perception that existing approaches are the reason for past achievements.

If everything appears to be working, why should leaders invest time and energy in making improvements? Why aren't they experiencing the pain that would motivate change?

The reality is they often don't—at least not until a significant event occurs or until internal or external forces dramatically shift the business landscape.

## Who This Book Is For

If you are responsible for leading a team that solves problems, generates ideas, innovates, makes critical business decisions, analyzes data, engages with customers, or implements strategies to drive growth—and you feel the frustration and concern that come with these responsibilities—then keep reading.

If you sense that your teams struggle with clear, effective communication or your team members stay confined to their silos without proactively sharing information, insights, or ideas, this book is for you.

*Transform or Die* is designed for leaders who aspire to achieve greater growth, scalability, and a stronger leadership pipeline. The tools and techniques inside will help improve creativity and productivity, leading to better team outcomes.

Whether you are a CEO, President, General Manager, Senior Vice President of a division, Vice President of a business unit, or the leader of knowledge worker teams, this book will show you how to unlock and harness the full potential within every team member during and beyond the current transformative moment in business history.

Within these pages, you will find essential context, a proven framework, practical examples, actionable tools, and exercises to enhance both human-to-human and Human+AI communication and collaboration. This book will guide you in turning your frustrations into positive, measurable results by providing a framework and roadmap for creating a culture where human and AI strengths are fully valued—and where collaboration becomes your ultimate competitive advantage.

## The Problems This Book Solves

**Within these pages, you'll discover frameworks and tools to help you:**

1. Transform AI adoption into measurable business impact by moving your workforce from simply experimenting with AI to delivering tangible bottom-line results.

2. Overcome ineffective collaboration by increasing your teams' effectiveness (with or without AI) in ways proven to boost margins, profits, and growth.

3. Avoid costly AI missteps by leading your teams with clarity so you don't waste millions on pilots that get stuck in the "purgatory bucket."

4   Eliminate uncertainty in AI investments by identifying high-impact initiatives and calculating ROI before committing significant time and money to resource projects.

## A New Model for Team Success: Human+AI Collaboration

Introducing new behavioral practices, such as round-robin brainstorming, silent idea generation, and role-playing exercises, can encourage teams to open up and share diverse perspectives. However, to transform business outcomes, the winners of tomorrow are rethinking their workflows and utilizing AI as a new strategic knowledge partner. **Human with Machine:** to create solutions neither alone could produce.

For this level of result, leaders need to realize that success ahead is not about the tech, but about the quality of the team dynamics, their innovative thinking, and decision making with AI as a new power tool.

The unique human strengths of empathy, creativity, and ethical judgment blend seamlessly with AI's capabilities, enhancing power and speed in data processing, analysis, and automation of tasks. Together, Human+AI Collaboration forms a new foundation for transformational business success.

## Your Next Steps

After four decades of working with leaders across diverse industries facing challenges similar to yours, I've learned that effective team collaboration extends far beyond having the right people on board or the latest technology at your fingertips—it's about practicing and cultivating the right team behaviors for everyone to thrive.

McKinsey's article "Go, teams: When teams get healthier, the whole organization benefits" (October 2024) confirms the value of investing in teams. The chapters ahead will guide you through a practical, step-by-step approach to help organizations of all sizes transform and enable their teams to work better together. You'll find proven frameworks, real-world examples, and actionable techniques you can put into practice immediately.

Now is the time for leaders to move beyond outdated notions of top-down, command-and-control leadership to coaching, mentoring, and leadership guidance for the exciting and unknown new future—by embracing the immense potential of human–AI partnerships.

Turn the page—and let's begin.

Given the rapid changes and evolution of AI Technologies, as a thank you for your time and readership, I am providing all readers access to Transform or Die as a living, consistently updated reference resource where you will find the latest information about AI tools and platforms, practice prompts, easy-to-use scorecards, and self-directing training programs for your teams.

Use this QR code or visit RussellMKern.com/transform to access the resources.

# Chapter 2

# When Your Teams Are Not Working Like Teams[6]

---

*"Leadership is about setting a vision and empowering your team to achieve it, especially during times of uncertainty."*
~ Jane Fraser, CEO of Citigroup

---

If you find yourself searching for answers to the following questions (or feeling frustration because of them), this book was written for you:

- Why don't our leaders spend more time collaborating on moving the big rocks of value creation rather than merely optimizing at the margins?

- Why don't my teams function like a cohesive unit?

- Why aren't my leaders truly leading?

---

[6] From multiple conversations with Sean Riches, CEO of Erskine Nash, Leadership Development Consultancy U.K. https://www.erskinenash.co.uk/team

- Why aren't my leaders developing their managers' coaching abilities to grow team members?

- Where are the great ideas, innovations, and strategies we need to achieve our strategic objectives?

## The Three Falsehoods of Leading Great Teams in the AI Age

### Falsehood #1: Get the Right People in the Right Role on the Team

In his influential book *Traction: Get a Grip on Your Business*, author Gino Wickman—creator of the Entrepreneurial Operating System®—asserts that "if you have the right people in the right roles and their roles are clear, then the team will be highly effective."

From my personal experience, forming and developing a high-performance team—a team that is highly successful in clear, authentic communication and feedback, effective collaboration, and high-performance, high-productivity behaviors—is far more complex than simply assigning the right people to the right positions.

Here's one reason why:

Our extraordinary 3-pound brains contain over 170 billion cells, including neurons, the primary signaling units that transmit information throughout the entire body. Our brains are not machines! We are living, complex, ever-changing, aging, knowledge-gaining organisms. Our brains are comprised of neurons that form nearly 100,000 miles[7] of electrical wiring, firing signals rapidly between 1 and 120 meters per second, while consuming just 20 watts of energy, all in a water-rich environment. When you reflect on that fact, you realize how unique, sophisticated, and magnificent every person you bring onto a team is. Our brains automatically respond to everything we see, hear, touch, taste, and smell. Meanwhile, our subconscious effortlessly keeps us alive, regulating our heartbeat, breathing, and digestion, and restoring our systems during sleep.

Still, after 7 decades of organizational behavior research, how humans think, feel, and behave within small groups, teams, and social tribes remains unknown to organizational psychologists. Unlike computers built from wires and circuits, humans bring evolved unconscious survival behaviors and complex emotional responses to teams. Even when individuals have the right experience and proven success, you can't assume that simply assembling a group of high performers will instantly create a top-performing team. Sports teams after sports teams spend billions to buy the best talent, only to find that human frailties, such as ego, insecurity, pride, and self-focus, fail to deliver championships.

---

[7]   Olaf Sporns. Networks of the Brain. Cambridge, Massachusetts, The Mit Press, 2016.

## Falsehood #2: Great Talent is Highly Adaptable to Technology Change

Over recent decades, the workplace has embraced transformative technologies, including copiers, fax machines, servers, desktop computers, the internet, websites, and voice recognition. In every case, experimentation, practice, and pilot projects have been the essential stepping stones to proving use cases, enhancing skill proficiency, and ultimately driving widespread technology adoption. However, what is not talked about is the adaptation and acceptance of the changes brought about by the technology.

As a global society, we have unlocked access to one of the most powerful and virtually limitless knowledge systems: artificial intelligence. We are seeing how AI can act as a human assistant, a librarian, and a supercharged strategic ideation partner, and become the ideal digital worker who never sleeps, eats, or complains.

Yes, this technology is capable of detecting patterns far faster than humans and recommending the next actions within seconds. When used thoughtfully, AI enables individuals and teams to work smarter and faster, spark more ideas, and generate innovative solutions beyond anything we could have imagined.

Here's the problem when it comes to leading teams. Humans are terrible at seeing and adapting to change. Our 3-pound brains crave stability, consistency, known patterns, and low survival threats. We evolved on flat plains over hundreds of thousands of years. We are terrible at seeing and accepting logarithmic change and transformations. Which means most team members are currently struggling with accepting the change before us. AI technologies and climate change are here, real, and pose threats while creating opportunities, requiring team members to shift their mindset so they can feel a sense of stability and achieve a positive outcome.

What this means for team leaders is an increased need to demonstrate compassion and patience.

The nature of work and the skills required evolved dramatically when society transitioned from horse-drawn carriages to automobiles in just 13 years, resulting in both job loss and the creation of new job opportunities. Today, leaders will be best served by understanding the AI transformations from a similar perspective, but at a speed of change that is 10 times faster.

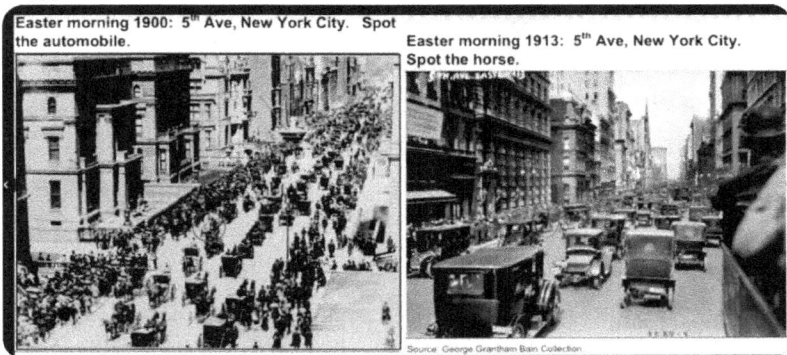

Easter morning 1900: 5th Ave, New York City. Spot the automobile.

Easter morning 1913: 5th Ave, New York City. Spot the horse.

Source: George Grantham Bain Collection

Which is why there should be no surprise when 50% of today's workforce is afraid of losing their jobs to AI and is resistant to fluency and adoption.

## Falsehood #3: Teams will Quickly Increase Their Productivity and Capabilities with AI

Learning how to write prompts into open AI platforms like ChatGPT, Claude, Perplexity, Gemini, Poppy, or DeepSeek will not automatically produce "magic" for any business leader or team. In fact, during the early years of AI start-ups, many CEOs have been disappointed by failing to see the bottom-line impact instead of the huge projected results surrounding AI hype.

**Why does it fail?**

Evidence from multiple case studies makes it clear: AI's promise and the gains it will deliver to business are real. However, realizing this value requires leaders to dedicate time and resources to strengthening one often-overlooked but essential skill: Collaboration.[8]

My experience and the research show that developing and elevating effective human-to-human collaboration is a leader's first priority. Then, leaders can develop the unique skill of humans working in combination with AI by using an iterative, questioning approach to see new perspectives and address unknowns.

In addition, without developing the foundational skill of AI fluency, AI investments will continually face roadblocks to realizing the transformative power that AI brings to business,

---

[8]    Sands, Dr Molly. "AI Collaboration Report: "Using" AI Is Not Enough - Here's What Your Organization Is Missing - Work Life by Atlassian." Work Life by Atlassian, 19 Nov. 2024, www.atlassian.com/blog/productivity/ai-collaboration-report.

stemming from a language and mindset barrier between operational units.

Compounding this falsehood are existing hierarchical organizational structures, entrenched silos, established cultural norms, the constant reshuffling of teams, and the relentless pressure to do more with less, all of which directly contribute to unclear communications and ineffective collaboration.

## The Three Challenges of Transforming Teams into Human+AI Collaborators

### Challenge 1: Building Meaningful AI Fluency and Proficiency Across Teams

When Sarah joined MidMarket Solutions as Chief Learning Officer, she encountered a fragmented landscape of AI initiatives scattered across different departments. Some teams were tentatively experimenting with basic AI tools, while others avoided them altogether. During her listening tour, she uncovered a troubling pattern: even in departments actively using AI, there was no shared understanding of the tools' true capabilities or limitations.

"I thought AI was just for automating reports," confessed one operations manager. "I had no idea it could help us with strategy."

AI literacy extends far beyond simply knowing how to prompt an open-source chatbot, use Co-Pilot to summarize a report, fix an email, or generate images in Midjourney. AI literacy requires a holistic understanding, including:

- **Understand AI:** Starting with what large language models are and how AI is trained means knowing both its core strengths and limits, as well as the many ready-made and custom AI tools that can deliver real value for specific tasks.

- **Know your DATA!** Understanding what data you will be pulling from - internal data, WWW, company data, or what data the AI was trained on, if you are using a vendor's AI tool, etc. This is critical; otherwise, the outcomes may not achieve your goal—especially if you are trying to create efficiency in an internal process versus ideation from an external perspective.

- Have the ability to discern where AI collaboration genuinely adds value versus when it introduces unnecessary complexity.

- Gain the  skill to identify and start with a smaller pilot that has a high probability of success and a meaningful result.

- Ensure you have a solid grasp of organizational ethics and governance to ensure responsible and ethical AI use.

Without foundational knowledge and comprehensive, hands-on learning, teams often remain stuck at a superficial level with AI, either failing to fully leverage its capabilities or over-relying on it in areas where human judgment is irreplaceable.

Leaders must begin transforming their teams by first investing in developing their own AI literacy and that of those around them. AI fluency and proficiency are not achieved through any single training event, workshop, or providing individuals with Udemy or LinkedIn Learning for self-study. Adult learning is most effective when it is on-the-job, hands-on, and directly relevant to the work.

Consequently, leaders will want to encourage team hackathons, safe experimentation, and peer-to-peer mentoring. Identifying ambassadors who actively engage skeptics, turning them into advocates for transformation, and sharing wins broadly is key to success. Leaders will also want to support and empower ambassadors to hold office hours, provide on-call support, and create libraries of applications and prompts tailored to specific functional areas, thereby preventing teams from having to reinvent the wheel.

## Challenge 2: Navigating the Fear Factor

The conference room fell silent as Carlos, CEO of a 600-person manufacturing firm, announced the launch of their new AI initiative. Despite his reassurances that AI would augment rather than replace jobs, a palpable sense of anxiety filled the room. One veteran engineer later confided, "Every article I read talks about AI taking jobs. How am I supposed to feel enthusiastic about something that might make me obsolete?"

This fear typically manifests in three distinct ways:

- **Existential anxiety:** Concerns about job security and professional identity

- **Competence concerns:** Worries about keeping pace with new skill requirements

- **Loss of agency:** Fear that human judgment will be devalued or overridden

Our negative emotions create a powerful undercurrent of resistance that can undermine even the most carefully designed human-AI collaboration strategies. What many leaders overlook is that this resistance is not simply about preserving the status quo; it reflects a deeper desire to protect human dignity and purpose in the workplace. Change is inherently difficult for humans, who have evolved to seek certainty and stability. To support new types of team collaborations with AI, leaders must acknowledge that fears—such as the loss of jobs or being replaced by digital workers—are genuine concerns that require open recognition and thoughtful response.

The most effective way to overcome these fears is not to dismiss them but to address them directly. Consider Salesforce's Agentforce platform[9], which supports 9,000 help desk agents. While the technology delivers answers 16 times faster with 75% greater accuracy, Salesforce recognized the need to redefine success metrics and career paths for their

---

[9]   Tamayo, Jorge, et al. "Reskilling in the Age of AI." Harvard Business Review, Harvard Business Review, 1 Sept. 2023, hbr.org/2023/09/reskilling-in-the-age-of-ai.

human employees. They shifted performance evaluations toward complex problem-solving, relationship management, and system optimization—areas where human expertise remains indispensable.

As AI technology continues to advance toward Agentic AI and Agentic Generative Intelligence, workforce anxiety is likely to intensify. The prospect of a $12 to $15 trillion digital workforce evolving alongside human teams raises profound questions for everyone about the future role of work.

Yet, it is essential to remember that humans possess irreplaceable qualities. Our curiosity, contextual judgment, and innate drive for self-determination provide unique advantages that no algorithm can replicate.

## Challenge 3: Closing the Team Leaders' Knowledge Gap

> Jamie, a division director at a financial services firm, captured this challenge perfectly: "I know we need to integrate AI into our workflow, but I have no roadmap. How do I coach my team through a transition that I barely understand myself? What does effective human-AI collaboration even look like?"

Many leaders feel significant pressure to transform the way their teams work, but often lack clear guidance on how to start. These knowledge gaps can hinder their ability to effectively support, invest in, coach, and develop their teams, thereby harnessing the unique strengths of both humans and AI. Too frequently, senior leaders delegate responsibility for team development to HR or learning departments, instead of recognizing that teams—one of the company's most valuable assets—benefit greatly from seeing leaders actively engage as teachers, coaches, and mentors.

This leadership knowledge gap typically emerges in three critical areas:

- **Strategic confusion:** Uncertainty about which processes and functions should be prioritized for AI integration

- **Development blind spots:** Lack of frameworks for building essential human capabilities alongside AI adoption

- **Cultural misalignment:** Failure to adapt team norms, incentives, and collaboration practices to a human–AI environment

The most successful leaders bridge this gap by actively engaging in the learning process themselves, not merely sponsoring it from the sidelines. They experiment alongside their teams, ask probing questions that uncover both opportunities and potential pitfalls, and foster a psychological safety that enables open discussions about successes and challenges.

Organizations that excel in human+AI collaboration understand that while technology implementation has become increasingly straightforward, integrating it into human systems demands careful and intentional effort. The tools are accessible; it is the alignment with human behavior and leadership practices that determines success.

## Moving Beyond the Challenges

Simply putting the right people in the right roles is only the starting point for achieving effective communication, collaboration, and teamwork in the AI era. Further, leaders understand the natural fear and reluctance to change, which is why they support the use of transformational leaders. Ultimately, they acknowledge that not everyone desires to work in a team. There are different working styles and comfort zones. Those leaders who possess high levels of emotional intelligence understand how to best weave a cohesive team through the use of various team-building strategies.

McKinsey reported in its article "Go, Teams: When teams get healthier, the whole organization benefits" that the four drivers of high-performing teams are: having high trust that everyone will do their job, being effective at communications, supporting innovative thinking, and enabling individuals to use their judgment.

These four characteristics explain nearly 75% of the difference between low and high performing teams, which can be measured by three key outcomes:

1. The team is efficient, productive, and meets its deadlines, accomplishes its objectives, and delights stakeholders and customers.

2. The innovation within the team delivers long-term organizational value.

Human+AI collaboration requires a fundamental rethinking of what teamwork means. It calls for leaders who can:

- Foster psychological safety while guiding teams through the natural fears that come with change.
- Establish clear boundaries that distinguish human decision-making from AI-supported decisions.
- Create a common language and a set of models to support AI collaboration.
- Design and implement consistent, hands-on, real-world practice sessions that develop the skills that bring out the best of human and machine capabilities when working together.

The organizations that are thriving with AI are not those with the most sophisticated tools, but rather those with the most cohesive processes to enable their human workforce to do what humans do best, while developing the skills to have their AI partners excel at what computers, algorithms, and models can do.

AI-first companies are already surpassing slower adopters by embedding AI into their core products, services, and processes. Progressing rapidly from individual productivity gains and pilot projects to ultimately reinventing entire workflows and offerings. The speed of this progression hinges on leadership.[10]

---

[10] "Succeeding in the Digital Age Why AI-First Leadership Is Essential." Harvard Business Publishing.

I've included detailed Case Studies of successful Human+AI teams inside the online living resource this book directs you to. For those who wish to learn more, I've also included the story of Ryan and TEDx business building using Agentic AI.

---

*"We are completely convinced the consequences will be extraordinary and possibly as transformational as some of the major technological inventions of the past several hundred years: … electricity, computing, and the Internet."*
—Jamie Dimon, JPMorgan Chase CEO

---

## It's Time to Act

Today's leaders must radically challenge themselves to rethink what it means to build, develop, train, and lead effective teams in the age of AI. Modern leadership requires shifting focus from simply managing tasks and outputs to intentionally cultivating environments where strategy, ideation, creativity, collaboration, and innovation are prioritized.

The world of work is undergoing seismic transformations:

- Millennials and Gen Z place a premium on purpose, equity, and flexibility.
- AI and digital technologies are reshaping workflows and team dynamics.
- The pace of change continues to accelerate, with the lifespans of S&P 500 companies shrinking at unprecedented rates.

- Teams must develop resilience, agility, and readiness to navigate uncertainty, from global crises to shifting cultural expectations.

Yet amid all this change, the constant is humanity. Every team member brings unique experiences, motivations, and ways of thinking. Trust, clear communication, and psychological safety to speak up, ideate, debate, coach, and mentor one another remain the bedrock of high-performing teams and organizations—and they are more critical than ever as we integrate AI into how we work.

---

*"Leaders need to begin their journey now to help their teams not be just agile and resilient enough to survive but to flourish as one team—a team that thrives in the face of complexity, uncertainty, and pressure by leveraging the vast knowledge that AI brings to all."*
—Russell M. Kern, CEO, Kern and Partners

---

The time to act is now. Businesses that fail to adapt will find themselves left behind in an increasingly competitive and knowledge-driven economy. For leaders eager to make meaningful progress, this book offers the mindset, strategies, and tools necessary to cultivate effective human-AI collaboration behaviors across every team in their organization.

This is more than a call for change—it is an invitation to lead the way. Today's leadership demands more than simply achieving objectives; it requires building a lasting legacy of collaboration, innovation, and sustainable success, all powered by a new era of human-assistive technologies.

Let's keep going.

Given the rapid changes and evolution of AI Technologies, as a thank you for your time and readership, I am providing all readers access to Transform or Die as a living, consistently updated reference resource where you will find the latest information about AI tools and platforms, practice prompts, easy-to-use scorecards, and self-directing training programs for your teams.

For this chapter, you will want to check out these resources:

- AI Literacy Checklist and Resource Library
- Managing AI-Related Change Anxiety
- Leadership Self-Assessment Tool for Human+AI Readiness

Use this QR code or visit RussellMKern.com/transform to access the resources.

# Chapter 3

# Russell Kern's History, Experience, and Qualifications

## Answers From Ashes

In January 2025, during the devastating Los Angeles fires, I found the answers I had been seeking for months to guide my consulting and training business—answers that would ultimately shape the direction and flow of this book.

As my wife and I stepped outside our home, winds whipped at over 70 miles per hour, and a massive black plume of smoke rose above us. We were told to evacuate immediately.

In less than half an hour, we loaded our three horses into the trailer, secured our two dogs, Penny and Teddy, and gathered what we considered essential. We were fortunate to return home the next day to find everything intact, thanks to the heroic efforts of the LA fire crews. While the fire took so much from so many, it unexpectedly gave me a gift: the gift of reflection.

*Without power for almost a week, I learned to be more present and truly pay attention to the things around me.*

During this time, the answers to several questions about the next chapter of my life became clear:

- What value do I have to share with others?

- What is my mission at this stage in my life?

- What are my passions, expertise, messages, and authentic perspective that will help others move forward?

- How can I bring together my 40+ years of experience as a serial entrepreneur, marketer, strategist, creative business leader, team dynamics researcher, neuroscience advocate, and student of AI technologies into one cohesive concept?

I will share the answers to these questions with you shortly. However, first, let me provide an overview of my training and expertise, along with a few key leadership lessons I learned both early in life and throughout my career.

## The Russell Kern Credential Summary

### Russell's Business Experience

I had an unforgettable 50-year journey as a serial entrepreneur, having launched and led ventures across retail, medical services, and consulting. I've experienced the huge highs of a buyout and have suffered the giant lows of the crushing blows of Covid. However, my longest and most transformative chapter began when I founded a small creative shop next to LAX airport. Though it started as two guys working two shifts, over the decades, as the agency's President, I was able to guide its growth into a national powerhouse. It ultimately attracted the attention of one of the largest global public agency holding companies, Omnicom, which acquired us in 2008.

I remained at the helm until 2023, steering the agency through 15 consecutive years of revenue growth, doubling our team from 200 to 400 members, and consistently delivering strong profit margins. Along the way, my team and I created award-winning campaigns and innovative solutions that fueled growth for iconic brands such as AT&T, AAA, American Express, Canon Solutions America, DirecTV, Blue Cross Blue Shield, HP, GSK, SAP, and many others.

It's important for me to share that my success would not have happened without the incredible teamwork of our senior leaders and my brilliant mentors.

Through this journey, I oversaw more than $1 billion in client marketing investments and orchestrated over 50,000 marketing tests, sharpening my expertise in both B2C and B2B strategic planning. This passion led me to author the S.U.R.E. Fire® strategic planning process and a book published by McGraw-Hill.

My commitment to lifelong learning has taken me through advanced studies in Neuroscience of Leadership at MIT,

Creative Leadership at IDEO, Appreciative Inquiry from Case Western Reserve University, and Learning & Development Operations at the Josh Bersin Academy, among others. I am certified in Working Genius and CliftonStrengths and have immersed myself in topics ranging from emotional intelligence and strategic planning to the neuroscience of persuasion and leading high-performance teams. As a member of SHRM, ATD, and the OD Network, I continue to explore the art of questioning, habit formation, and leadership, always seeking new ways to inspire teams and drive meaningful results.

## Lessons From My Youth

Three lessons from my childhood continue to shape my values as a leader to this day.

### Lesson #1: The Importance of Initiative

When I was ten years old, I attended sleep-away camp for the first time at Jameson Ranch Camp, located outside Bakersfield, California. On our very first night, all the campers gathered around as the founder, Mr. Rod Jameson, shared a story that would become a cherished camp tradition.

He told us about two campers: Michael and Karen. Both spent the same days at camp—they rode horses, swam, ate the same meals, and had the same counselors. Yet when camp ended, Michael went home sad, while Karen left with memories she'd cherish forever.

What made the difference? Mr. Jameson explained that Michael never got involved. He didn't raise his hand for activities and spent most of his time sitting on his bunk, waiting for something to happen. Karen, on the other hand, volunteered for everything and eagerly participated in as much as she could.

Mr. Jameson concluded his story with a simple lesson that has always stayed with me: you get out of camp what you put into it.

---

*You get out of life what you put into it.*

---

## Lesson #2: Teamwork Creates Lifelong Connections

I discovered that summer camp would have been an entirely different experience without the teamwork involved in washing dishes, weeding the garden, or rounding up our horses. Jameson Ranch Camp was built on the spirit of teamwork because on a working ranch, everyone has to pitch in or the ranch couldn't operate. While ranch work is never-ending, our teamwork transformed the chores of the day into pure joy.

I still carry with me the sense of camaraderie that grew between counselors and campers as we hiked, rode, swam, or even participated in arts and crafts. More than sixty years after my first night at camp, these experiences continue to shape my core values and leadership philosophy.

Me, age 17, as a CIT at Ranch Camp with my classic 70s hair.

There's joy in teamwork.

## Lesson #3: Every Person is Magnificent.

Learning to sell and manage a three push cart business, selling Jelly Bellies between my Sophomore and Junior years in college.

---

*"You can tell a lot about a fella's character by whether he picks out all of one color or just grabs a handful."*
—Ronald Reagan, President of the United States

---

Between my sophomore and junior years of college, I took a full-time job selling Jelly Bellies. This job taught me valuable lessons about packaging, distribution, sales, and employee management—skills essential for running a small retail business. Yet one lesson proved far more meaningful than all the rest.

What sets Jelly Bellies apart from other jellybeans is that their intense flavor is concentrated inside each bean rather than being blended into a thick sugar coating like traditional jellybeans. This unique process creates magical flavors that closely match their names, from Very Cherry and Root Beer to Tangerine and even Buttered Popcorn.

> The lesson I learned from my time selling Jelly Bellies was:
>
> The deliciousness, the wonderfulness, the magnificence of every Jelly Belly is INSIDE—in the heart of the bean.

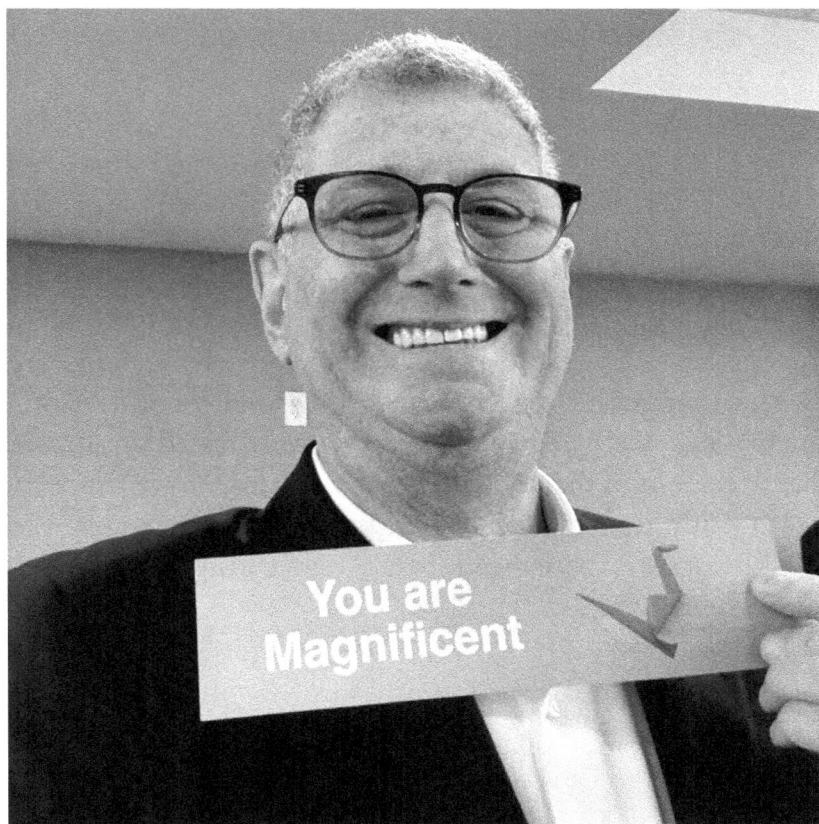

My favorite method of appreciation is sending this YOU ARE MAGNIFICENT selfie photo in a text with a special note.

This lesson extends far beyond candy: every human being carries their own wonderfulness and magnificence within their heart and soul. As leaders, it is our responsibility to recognize the greatness inside others—no matter how it presents itself—and to encourage it to flourish.

*Acknowledging a person's natural gifts, their magnificence, is one*
*of the greatest gifts one person can give to another.*

## Lessons from My Early Entrepreneurship Ventures

From a young age, I launched and grew a variety of businesses—some solo projects that remained small, and others that scaled into profitable, national organizations, such as my creative agency. Each venture provided invaluable, hands-on lessons that refined my expertise in creativity, strategy, team building, collaboration, and technology. Below, I've distilled those experiences into actionable insights for leaders and entrepreneurs.

### My Early Business Ventures and Key Lessons Learned

| | |
|---|---|
| Briarhills (mail-order executive gift company I started at the age of 21) | Lesson: Economy of scale matters—small businesses face higher costs and lack competitive pricing or marketing advantages. Packing and shipping 200 of anything is a lot of work! |
| Parsons and Kern (My first ad agency partnership with Renaissance creative Carl Parsons at 25 years old) | Lesson: Opportunities are often right next to you—our first big client came from the PR firm referral right across the hall. |
| Kern and Mathia (Second incarnation of the ad agency) | Lesson: Big offices don't win clients; investing in people does. |
| Kern Direct (Third agency, this time solo) | Lesson: Reactive leadership drives away valuable team members. |

## Leadership Blind Spots and Growth Areas

During my early years, I made several classic leadership mistakes that became pivotal turning points:

- I blamed others for business challenges instead of taking ownership.

- I didn't realize that great leadership involves acting as a growth coach for every team member.

- I lacked coaching skills and didn't know that coaching was something I could learn.

- Without leadership role models—coming from a family of medical professionals—I didn't even understand what being fired meant until my first job after college.

- I modeled outdated, hierarchical, command-and-control, bossy leadership styles due to a low self-awareness of my actions and their impact on others.

- I lacked knowledge on how to effectively lead and grow a profitable creative services business, which resulted in the loss of valuable team members and hindered the agency's growth.

---

*The "school of hard knocks" and support from mentors eventually helped me turn these blind spots into strengths.*

---

# Invaluable Business Lessons from Later Leadership Years

As my agency expanded—growing services and staff before ultimately being acquired by Omnicom—my leadership skills and approach evolved significantly. Here are the most impactful lessons I learned during those years:

### Work with an Incredible CFO
Surround yourself with senior team members who provide unfiltered advice and mentorship. I was fortunate to have one of the greatest and most forthright CFOs as a colleague.

### Invest in Leaders and Team Relationships
Help everyone identify and leverage their strengths. Everyone is magnificent if you look for it and celebrate it.

### Embrace vulnerability
As Brene Brown advocates, vulnerability is the gateway to deeper, authentic connections.

### Give Yourself the Gift of a Great Coach
Just like elite athletes, every leader benefits tremendously from skilled coaching.

### Create a Culture of Safety
The worst thing any CEO can do is surround themselves with YES people. Encourage and cultivate an environment where team members feel safe to share ideas, challenge one another, and engage with you without fear of reprisal.

## Encourage Experimentation
Great success arises from huge failures. Commit to testing early and often. Test ideas, sales strategies, scripts, and processes to drive continuous growth and profitability.

## Encourage Collaboration Across Silos
As organizations grow, cross-team communication can deteriorate. Make cross-silo collaboration a priority to strengthen engagement and alignment. Make time for cross-silo leaders to focus together on moving the big strategic rocks of the business, not just optimizing on the fringes of the business.

## Invest in Learning and Development
For decades, learning and development investments have been seen as a cost center because many leaders can't tie these investments to business outcomes. My experience is that this is exceptionally short-sighted management, given that one of your most expensive and valuable assets is your workforce. Investing in the growth of your workforce, deepening their skills, capabilities, and capacity, pays off in the form of reputation, engagement, passion, referrals, and customer satisfaction, which all translate into top and bottom-line growth. All businesses are run by people. Love and grow your people, and watch your business grow, too.

---

*These lessons form the foundation of my approach to leadership and team effectiveness. They continue to shape my work as a transformationalist, collaboration expert, strategist, mentor, coach, advisor, and author.*

---

## Why This Book? Why Now?

In 2023, a unique convergence of circumstances inspired me to embark on a new chapter: launching my consulting, coaching, and training business with the writing of this book, *TRANSFORM or DIE*. After four decades of leading my agency, it was time to pass the reins to the next generation. Yet my passion for leadership, creativity, strategic thinking, team dynamics, and coaching—all areas I have studied extensively—remained as strong as ever. Simultaneously, the rapid rise and adoption of artificial intelligence technologies opened up exciting new possibilities, a frontier that had never been able to scale before.

I am energized by AI's potential to serve as a strategic partner for nearly every knowledge worker, enhancing not only personal productivity but also expanding the human race's collective capacity to ideate, innovate, and collaborate with one another and technology in ways we once only imagined.

To fulfill this mission, I knew I needed to share not just my credentials and experiences but also develop a methodology for developing the skills of collaboration for a range of problem-solving in an easy-to-remember, credible, and practical way for the business world.

## Ending Where I Started

I want to close this chapter by sharing the answers to the questions I had been pondering long before the LA fires. These insights came to me in a flash, one morning as the fires were still burning outside my front door.

## What measurable value will you gain from my experience?

- Skip the expensive leadership lessons I learned the hard way. Avoid **the million-dollar mistakes** that cost me market share, good people, and competitive edge when my company needed it most. Learn how to sidestep these costly missteps to better focus your energy on what drives growth.
- Tap into four decades of building and scaling businesses where I've wrestled with the same challenges keeping you up at night: getting teams to innovate under pressure, breaking down silos that kill good ideas, and figuring out how to use new technology without breaking what already works.
- Get a practical system that actually improves how your people solve tough problems and create together—not another theory, but a framework I've tested and proven for years that helps teams think better, decide faster, and deliver great results that move the needle.

## What is my mission at this stage in my life?

I am deeply committed to helping leaders and their teams transform, thrive, and flourish through improved collaboration.

## What are my passions, expertise, messages, and authentic perspective that will help others move forward?

- I love helping leaders "make their workplaces, their teams, and every person they work with just a little better each day."

- Guiding leaders to drive business success by increasing team trust, communication clarity, and fostering creativity, with and without AI technologies.

**How can I bring together my 40+ years of experience as a serial entrepreneur, marketer, strategist, creative business leader, team dynamics researcher, neuroscience advocate, enthusiast, and student of AI technologies into one cohesive concept?**

- Create the K.E.R.N. Human+AI Collaboration Framework, a development process that helps CEOs gain business advantages from great human-to-human collaboration—something I have practiced for 40 years.
- Share why leaders need to transform their mindset about the profound and ongoing impact of AI on their business.
- Demonstrate how successful AI collaboration requires knowing when to utilize it, which tools to leverage, and how to harness its vast knowledge and computational power.
- Guide leaders to overcome their most common challenges by providing hands-on practice with human+AI collaboration across silos, using AI as their strategic knowledge partner.

## What's Next?

We are about to explore the strategic and financial benefits of high-performance teams that excel at collaboration.

You will be introduced to the concept of Human+AI Collaboration. Ahead lies a deeper understanding of AI as a team's strategic knowledge partner—a partner that can help teams ideate more broadly, accelerate creation, and expand thinking in new and unexpected ways. AI can support the development of strategic solutions and evaluate the impact of plans at speeds and with insights beyond what humans alone can achieve.

You will also learn about the K.E.R.N. Human+AI Collaboration Framework and how to apply it.

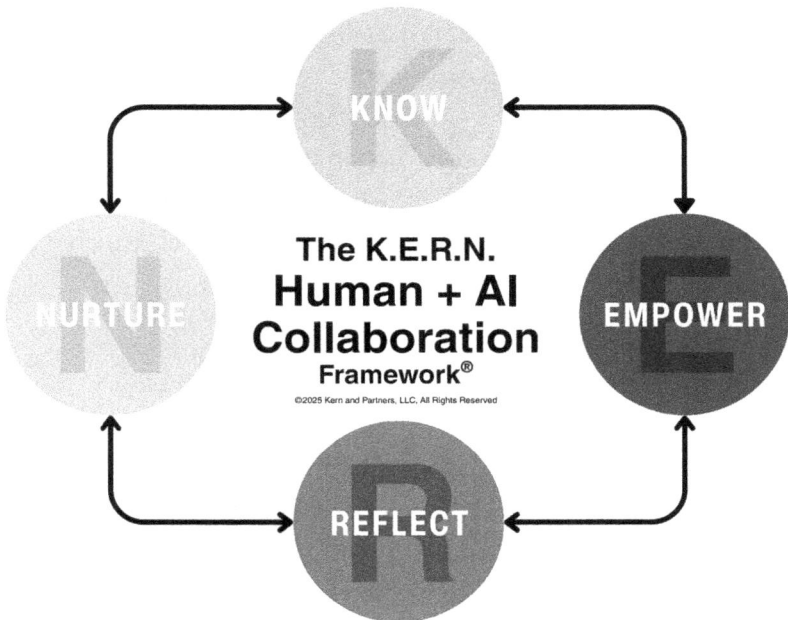

The K.E.R.N.
**Human + AI Collaboration** Framework®

©2025 Kern and Partners, LLC, All Rights Reserved

KNOW

EMPOWER

REFLECT

NURTURE

This four-part framework is captured in a simple, memorable acronym:

- **K** stands for **Know**—Yourself, others, and AI technologies.

- **E** stands for **Empower**—Exploration and experimentation.

- **R** stands for **Reflect**—on progress and results by taking a breath and stepping back.

- **N** stands for **Nurture**—lessons learned, positive progress, and the future success for all.

I hope you find the upcoming chapters enlightening and inspiring, and that they ignite your curiosity further to explore the powerful synergy between humans and AI. Enjoy your journey into the future of team collaboration.

Given the rapid changes and evolution of AI Technologies, as a thank you for your time and readership, I am providing all readers access to Transform or Die as a living, consistently updated reference resource where you will find the latest information about AI tools and platforms, practice prompts, easy-to-use scorecards, and self-directing training programs for your teams.

Use this QR code or visit RussellMKern.com/transform to access the resources.

# Chapter 4

# Incredible Team Accomplishments Begin with an Unwavering Commitment to Each Other

"This is an incredible honor for me to be standing here representing the Los Angeles Dodgers and this group of staff and players behind me," Kershaw said during his speech. "The 2024 Los Angeles Dodgers season is one that will go down in the history books.

President Donald Trump listens as Dodgers pitcher Clayton Kershaw speaks during a ceremony to honor the Major League Baseball 2024 World Series Champion team in the East Room of the White House, Monday, April 7, 2025, in Washington. (AP Photo/Evan Vucci)

"Always holding a special place in the hearts of myself, and those there with me as well as millions of Dodger fans around the world. This organization exemplifies what it means to come together as one for a greater purpose and represent something so much bigger than themselves.

"The selflessness and humility that each one of these players and staff has shown over the last year is truly an inspiration. They have constantly played hurt, switched positions, and taken the ball to put the team first. As a spectator for our championship run last year, I was in awe of this group.

"Their unwavering confidence, coupled with the selfless pursuit for team excellence, was an inspiration. That is why I'm so grateful to get to speak today on their behalf, as I know none of them would say this about themselves.

"Moving forward, I hope the 2024 Dodgers can serve as an inspiration to many, like they were to me. Not just in sports, but in life, remembering to put others before ourselves. It moves a team and a society forward. Thank you for allowing me to speak today, because the story of the 2024 World Series champion Los Angeles Dodgers is a true joy to tell."

Clayton Kershaw

Kershaw spoke on behalf of his teammates and joined owner Mark Walter in presenting a Dodgers jersey with "Trump" and "47" to the United States President, Mr. Donald J. Trump.

## Getting Great Ideas to Fly: The Collaboration at Kitty Hawk

The wind howled across the desolate sand dunes of North Carolina's Outer Banks, whipping grains of sand against the small wooden shed that served as both workshop and living quarters. Inside, two brothers hunched over a damaged wing strut, their faces illuminated by lamplight.

"Dammit, Orv," Wilbur muttered, running a calloused hand through his hair. "That's the third time we've snapped this junction. Maybe Chanute was right about the curvature."

Orville didn't respond immediately. He turned the broken piece in his hands, examining the fracture. Outside, the December chill penetrated their makeshift camp, a far cry from their bicycle shop back in Dayton. Three years of coming to this godforsaken beach, and still no sustained flight.

"What if," Orville finally said, reaching for his notebook, "we reinforce it here, but change the attachment point?" He sketched rapidly, the pencil smudging under his thumb. "See? Less torque when the wind gusts."

This scene, repeated hundreds of times with different problems and solutions, captures the essence of what made the Wright brothers' collaboration at Kitty Hawk extraordinary. It wasn't just that they succeeded; it was how they repeatedly failed together, yet kept going.

Few people remember that before December 17, 1903, the brothers had already suffered countless setbacks. Their 1901 glider performed so poorly that Wilbur declared they might be thirty years away from flight. Their calculations, borrowed from established aeronautical authorities, proved dangerously wrong. Storms destroyed their first camp at Kitty Hawk. They battled mosquitoes, isolation, and the ridicule of newspapers that called them "flying fools."

Though the Wright Brothers grew up in Ohio, they found the perfect place for their flying machine experiments in Kitty Hawk, North Carolina. The brothers pored over weather records before determining that North Carolina would suit their needs. The first flight, on Dec. 17, 1903, lasted just 12 seconds and covered 120 feet. By the end of the day, the world's first airplane stayed in the air for nearly one minute. *Public Domain / Wikimedia Commons*

Through it all, they maintained a partnership unlike any other. When arguing over technical problems (which happened daily), they would switch positions, each brother arguing the other's viewpoint to test its merit. This wasn't just a debate; it was intellectual humility in action. They understood that ego was the enemy of innovation.

Their complementary personalities created a balance that neither could have achieved alone. Wilbur—more cerebral and strategic—provided theoretical direction. Orville—mechanically gifted and detail-oriented—translated concepts into reality. Both were stubborn as mules, but their stubbornness was directed at problems, never at each other.

The brothers' collaboration extended beyond their twosome. Back in Dayton, Charlie Taylor, their shop mechanic, built their custom engine when no manufacturer could meet their specifications. In Kitty Hawk, the Tate family did more than provide hospitality—they saved the brothers' lives during a brutal storm, helped drag their heavy machines across the sand, and connected them with locals who provided critical knowledge about the area's winds and terrain.

During one particularly frustrating week in 1902, after multiple crashes of their improved glider, Wilbur remarked to Orville, "It's not that I think we're going to fail. It's just that our path to success is longer and more winding than we imagined." That night, they sat by their campfire reexamining every assumption they'd made, questioning data that respected scientists had published as fact.

This willingness to challenge authority led to their breakthrough. Using a homemade wind tunnel constructed from a wooden box and hacksaw blades, they tested over 200 wing shapes, discovering that every published table of lift coefficients was wrong. Their meticulous records, filled with crossed-out calculations and revised numbers, reveal the painstaking collaboration that others weren't willing to undertake.

On that historic December morning, success didn't come easily. Their first attempt failed when the Flyer stalled and dropped to the sand. After repairs and adjustments—with a frigid wind making their hands so numb they could barely grip tools—they tried again. The famous first flight lasted just 12 seconds and covered only 120 feet, less distance than the wingspan of a modern passenger jet.

Yet in that brief moment, the result of thousands of hours of collaboration overcame what many considered impossible. The brothers didn't celebrate with grand speeches or champagne. Instead, Wilbur simply nodded to Orville and said, "We'll need to reinforce the front elevator for the next one." Then, they went back to work, making three more flights that day, each longer than the last.

The Wright brothers' achievement wasn't just technical; it represented the power of human relationships built on absolute trust, committed conflict over ideas (never personalities), and shared purpose that transcended individual recognition. When reporters later asked who deserved credit for the invention, neither brother would accept individual praise, with Wilbur famously responding, "I couldn't have done it without Orville, and he couldn't have done it without me."

Orville looks on as Wilbur prepares to take off with sister Katherine on her first flight, near Pau on February 15, 1909. (Courtesy C.V. Glines)

Their story reminds us that behind every great innovation lies not just brilliant minds, but minds that know how to connect, challenge, and complete each other. The miracle at Kitty Hawk wasn't just that humans finally flew—it was that two humans, through extraordinary collaboration, discovered together what neither could have found alone.

SpaceX and Elon Musk have consistently emphasized the importance of experimentation, learning from failure, and embracing risk as part of their innovation strategy. Musk has famously described the explosion of SpaceX's Starship rocket as a "rapid unscheduled disassembly," framing such events as opportunities for improvement rather than setbacks. Following a test flight that ended in an explosion, SpaceX stated, "With a test like this, success comes from what we learn, and today's test will help us improve Starship's reliability as SpaceX seeks to make life multi-planetary."

## When to Ideate? When to Collaborate?

I had the opportunity to speak with Joretha Johnson, President of Advanced Transformational Technologies and former president of the OD Network. Drawing on her extensive experience, including two decades at a leading global consumer products company, she shared valuable insights about how effective team leaders know when to direct their teams toward ideation and when to focus on collaboration. These two processes complement each other, and understanding when to employ each can significantly enhance team dynamics and problem-solving capabilities.

### Ideation: Use It When Pieces of the Solution Are Missing

*High-performing teams turn to ideation when gaps exist—whether in knowledge, resources, or clarity about the problem itself.*

When you are missing pieces for a solution, teams should use this time to explore possibilities and generate creative solutions. Engage your teams in brainstorming sessions using structured methods like S.C.A.M.P.E.R., which encourages unconventional thinking by taking an existing idea and applying techniques such as:

- Substitute
- Combine
- Adapt
- Modify
- Put to another use
- Eliminate
- Reverse

Ideation thrives on diverse perspectives and an environment that supports risk-taking without fear of judgment. Skilled facilitators are essential in guiding these sessions, striking a balance between focus and freedom to explore bold concepts. The objective is not to find immediate solutions but to generate a wealth of innovative ideas that can later be refined through collaboration.

Design firms, such as IDEO, the renowned product development firm, champion the importance of the ideation phase. They emphasize that success is driven by the quantity of ideas produced rather than the initial quality of each idea; refinement comes later in the process. During ideation, avoid editing ideas prematurely; instead, aim to gather as many ideas as possible from as many contributors as you can. This is why having strong AI fluency, which I refer to as power prompts, is helpful in generating additional ideas from various perspectives.

## Collaboration: Use It When All the Pieces Are in Place

*Collaboration thrives when a team has assembled the necessary resources, knowledge, and tools to effectively address a problem.*

In collaboration mode, the team's focus shifts to harnessing the collective intelligence and diverse perspectives of the team to efficiently execute impactful solutions. Team members work together to refine ideas and align strategies, creating clear pathways for seamless implementation.

Open communication, honest feedback, and candid debate put pressure on ideas, helping to uncover the most valuable solutions. Trust within relationships ensures that each member's expertise contributes fully to the shared goal. When everyone feels both compelled and safe to speak up, challenge ideas, and suggest new approaches, teams build a culture of mutual respect and shared accountability. This culture supports effective problem-solving while strengthening team cohesion and morale.

When teams collaborate effectively, they transform individual contributions into unified action, achieving results that are impossible to attain in isolation, just as demonstrated by breakthroughs like the Kitty Hawk or SpaceX.

Understanding the differences between ideation and collaboration—and knowing how to transition seamlessly from one to the other—ensures teams are prepared to tackle well-defined challenges or ambitious, mission-critical problems with both creativity and precision. Establishing this foundation is essential before introducing AI technologies as a new team member.

## Effective Collaboration Is Essential Given the Complexity and Interconnectedness of the World

Eric Beinhocker, Executive Director of the Institute for New Economic Thinking at the Oxford Martin School, notes that no country on Earth can produce an iPhone entirely on its own. No single nation, company, or team possesses all the knowledge, components, manufacturing capabilities, or raw materials required to create this remarkable device that fits in your pocket.

NASA highlights a similar level of interdependence: thousands of skilled scientists, engineers, and technicians from 309 universities, national labs, and companies across the US, Canada, and Europe contributed to the design, construction, testing, integration, launch, commissioning, and operation of the James Webb Space Telescope.

As Beinhocker explains, "When we can combine more complex knowledge and complex parts to solve complex problems, our quality of life soars. That is the most powerful engine for today's organizations: the division of knowledge and the collaboration required to make more complex things."

CORE CONCEPT: If a team cannot respectfully listen to each other and collaborate effectively, human-to-human, adding AI technologies as collaborative, strategic, or analytical partners will not improve team dynamics; rather, it will likely increase conflict, uncertainty, and stress.

That's why the essential first step toward successful human+AI collaboration is to build strong human-to-human collaboration skills using real work experiences.

## Why Develop Team Leaders Who Inspire Collaboration and Creativity?

Deloitte found that while 94% of organizations have deep expertise within their teams, only 12% effectively harness this collective intelligence through collaboration. This gap represents a significant opportunity.

In high-pressure environments—such as emergency rooms or crisis response teams—collaboration, also known as "teaming," becomes an essential skill. Organizations that master teaming don't merely survive complexity; they thrive in it.

The benefits of developing leaders who champion collaboration and creativity are undeniable:

- 73% of employees in collaborative workplaces are more engaged (Harvard Business Review)
- 43% faster problem-solving (McKinsey)

- 50 minutes saved per day, on average, through better collaboration (McKinsey)
- 60% increase in employee innovation (Deloitte)
- 67% more knowledge sharing across departments (Forbes)
- 72% faster identification and replication of best practices

Organizations that cultivate a culture of collaboration aren't just enhancing performance—they're transforming how they operate, adapt, and innovate. The result? A powerful competitive advantage and a workplace where everyone's wisdom truly matters.

## The Impact of Great Collaborative Cultures

When team members trust one another, they share information openly, make faster and more informed decisions, and achieve greater results together. Highly collaborative teams respond swiftly to change, attract top talent, and build enduring competitive advantages. The cycle of trust and collaboration fuels continuous improvement and organizational excellence.

### This Can Be Your Team

Imagine stepping into your office or joining a video call and immediately feeling energized, confident that your team has your back. Every member is focused, engaged, and eager to contribute, building on each other's ideas without hesitation. Leadership shifts naturally to whoever's expertise is needed, and open, honest conversations are the norm.

You're not alone—your team celebrates wins together, tackles challenges directly, and creates lasting value. This isn't a fantasy; it's the reality for leaders who inspire and cultivate true collaboration. When teams operate this way, success becomes not just possible, but inevitable.

---

*Human-to-human collaboration is the #1 essential component of effective team performance.*

---

Understanding the difference between ideation and collaboration is essential before introducing new technologies. Teams must first master working together by sharing knowledge, challenging ideas, and building trust to speak honestly, with compassion, before AI can truly become an effective partner.

## Human Relationships Will Always Be the Foundation for Collaboration Success

Strong relationships are at the heart of great collaboration, built on trust, support, growth, encouragement, vulnerability, and genuine connection. When leaders coach their teams by providing clear direction, setting specific goals, reinforcing team and organizational purpose, and establishing shared standards and norms, everyone thrives.

*If you want your team members to take your business to where you envision your success to be, you must start by investing in relationships.*[11]

**One of the best leaders of a championship sports team is Dave Robertson of the 2024 World Series Dodgers.**

"I think part of it is that's who I am. I think I'm a very positive, welcoming guy. I like to learn more about people... learn about people. I played the game, I know how hard the game is.

"I really appreciate that the game is about the players, but it's bigger than all of us. And I don't hesitate in letting the players know that, albeit it's still about them in the here and now. I just try to be their biggest champion."

"I just try to be sensitive to kind of where they're at in the moment personally, trying to get the best out of them."

"I communicate with each guy differently, so there's a lot of conversations."

"But I think the ultimate goal has got to be about our ballclub."

---

[11] Katzenbach, Jon R, and Douglas K Smith. The Wisdom of Teams : Creating the High-Performance Organization. Boston, Harvard Business Review Press, 1993, p. 111

> "I just try to make it about them, not about me, and about the Dodgers. But we have a lot of hard conversations, fun conversations. That's probably the crux."
>
> ~ Words of Managerial Wisdom from Dave Roberts, Dodgers MLB - Head Coach[12]

Roberts has the reputation of being a players' manager and someone who is very personable and approachable. Being a manager who is a former player carries significant weight in the clubhouse, as it helps him command respect. However, it also enables Roberts to relate to his players on a more personal level, which goes a long way.

While those characteristics and values may be an intrinsic part of him, Roberts has drawn inspiration from various sources to shape his managerial style. He has even taken aspects of managers he admired from his playing days.

One of those inspirational managers is Terry Francona, who managed the Boston Red Sox in 2004.

Francona believes that extraordinary communication is essential for developing a successful organization. This is reflected in how Roberts maintains a positive and welcoming approach, recognizing that the game is ultimately about the players.

---

12 Ibarra, Sebastian Abdón. "Dave Roberts Explains Players' Manager Reputation." *Dodger Blue*, 26 Jan. 2025, dodgerblue.com/dave-roberts-communicates-with-each-guy-differently-in-dodgers-clubhouse/2025/01/25/. Accessed 11 Feb. 2025.

It is understood that a great deal of coaching adaptation occurs throughout a single regular season. And, for teams that reach the playoffs, the conversations had with each player and the team as a whole can take on a new tone and require a new strategy to help them rise to that level of competition.

---

*"I feel some players are in a good place.*

*Some players, I think, I need more out of.*

*Some people are starting to feel a little bit of the anxiety.*

*Some people might need a little bit of reassurance, encouragement."*
—Dave Roberts

---

Roberts specifically identifies what each player needs from him in order for his team to be playing at the highest level during the most important part of the season.

## The Famous Words of Winners: "This Team Has Always Got My Back"

**How strong is this feeling and the daily actions within your teams?**

Watch almost any post-game interview on the field or court, and when the interviewer asks a winning team member how they feel in that moment, you'll often hear the phrase, "This team always has my back." When team members trust and rely on one another, they form bonds filled with positive emotional energy that drive success, even when facing significant challenges.

One of the greatest factors influencing business success is building this kind of trust. Having each other's backs stems from a leader's ability to create a culture where people feel safe to experiment, discuss, debate, innovate, and take calculated risks that elevate team performance.

That said, sports analogies aren't always a perfect fit for business teams. Unlike in sports, business teams must navigate ongoing ambiguity, evolving roles, and complex interpersonal dynamics, requiring leaders to intentionally foster environments where collaboration can truly flourish.

### New York Liberty Celebrate WNBA Championship Victory on 'GMA'[13]

The New York Liberty celebrated their historic WNBA championship win during a special "Good Morning America" appearance in Times Square. The team clinched its first title by defeating the Minnesota Lynx in a thrilling Game 5 of the series, marking a monumental moment in franchise history.

Brianna Stewart, who played a crucial role in the championship victory, emphasized the importance of teamwork and trust. **"I just know my team has my back.** Whether it was in Game 1 or missing two shots earlier, it was just a moment, an opportunity that I couldn't let pass by. When your team trusts and believes in you, it gives you even more confidence."

Sabrina Ionescu shared her thoughts on the team's journey, saying, "To see how far we've come in such a short

13   Grant, Shawn. "New York Liberty Celebrate WNBA Championship Victory on "GMA."" *Thesource.com*, 22 Oct. 2024, thesource.com/2024/10/22/new-york-liberty-celebrate-wnba-championship-victory-on-gma/. Accessed 11 Feb. 2025.

amount of time—doing this for each other and for the organization that has believed in us since the beginning."

This sports team's success results from hundreds of habits, shared beliefs, working norms, and acceptable behaviors, combined with countless hours of practice and a relentless focus on the team's mission, regardless of the score in any given game.

For business leaders striving to inspire cultures of high collaboration and dynamic teaming within their organizations, I cannot stress enough that while AI technologies can help teams work faster, smarter, and more creatively, it is not the technology alone that makes this possible.

Ultimately, it is an inspirational leader's example—their collaborative behaviors, energy, passion, and commitment to building strong relationships—that creates the conditions for positive and high-performing team dynamics.

## What Have We Discovered?

This chapter illustrates the transformative power of collaboration and ideation in driving organizational success.

Through historical examples, such as the Wright brothers' groundbreaking achievements at Kitty Hawk, and modern innovation leaders like SpaceX, we see how trust, open communication, and a shared vision are critical for overcoming challenges and achieving ambitious goals.

Collaboration thrives when teams leverage their collective intelligence, while ideation sparks creativity to address gaps and uncertainties. Together, these complementary processes form the foundation for adaptive, high-performing teams capable of navigating complexity with precision.

We also see real-world evidence that fostering a culture of collaboration is not just a strategic advantage but essential for business success in today's interconnected world. From the intricate teamwork behind the James Webb Space Telescope to the collaborative efforts required to create everyday marvels like the iPhone, it is clear that no single entity can succeed in isolation.

Leaders play a pivotal role in cultivating this culture by modeling collaborative behaviors, breaking down silos, and aligning team dynamics with organizational objectives. By doing so, they unlock the full potential of their teams, enabling innovation, efficiency, and a sustained competitive advantage. Ultimately, collaboration is not merely a tool—it is the engine of progress for thriving organizations.

Now, in Chapter 5, we'll explore how human+AI collaboration can empower your team, helping both your people and your business grow and thrive by making AI a valuable member of your team.

Given the rapid changes and evolution of AI Technologies, as a thank you for your time and readership, I am providing all readers access to Transform or Die as a living, consistently updated reference resource where you will find the latest information about AI tools and platforms, practice prompts, easy-to-use scorecards, and self-directing training programs for your teams.

For this chapter, consider reviewing the Team Collaboration Assessment, which provides customized insights to help you boost your results.

Use this QR code or visit RussellMKern.com/transform to access the resources.

# Chapter 5

# The Neuroscience of Human+AI Collaboration

Picture your team standing at the edge of a cliff, gazing out over a valley of opportunity. The path forward is uncertain, but it is navigable. Some teammates are eager to lead, others hesitate, and a few freeze.

With a leader having an understanding of how each person is processing a given situation, they can help the team move forward together.

This scene serves as a simple yet powerful illustration of how every team member's brains are wired to create different emotional responses to change, challenge, and collaboration. Recognizing differences is a constant within any team, and it is one of the keys to building and leading highly effective teams. Welcome to perhaps the most fascinating aspects of creating and coaching high-value teams: working with, rather than against, the unique human brain and its emotional responses from every team member.

While there are plenty of team leadership books on the importance of Emotional Intelligence, Understanding Work Styles such as (DiSC), there are few that discuss the neuroscience of team leadership. A little-known fact is that even your most enthusiastic team members have brains simultaneously wired for both collaboration and resistance to change. Nor will they reveal that your most innovative thinkers have neural pathways that can slam shut at the first sign of social discomfort. Even your highest performers, who could be your extroverts with a "driver" work style, may have brain pathways that inadvertently sabotage themselves and the very team cohesion they profess to value.

Understanding these dynamics is essential for all team leaders to foster effective collaboration and manage the human behaviors and dynamics within their teams. This often-unspoken factor shapes team members' behavior, guiding transformation while also contributing to resistance toward new processes and technologies.

## The Collaboration Paradox

Collaboration is the golden promise of every team and the dream of every CEO—yet, paradoxically, it's often the very thing that causes teams to stumble.

We gather our brightest minds, invest in cutting-edge tools and leaders, then encourage everyone to "work together," only to find the team mired in confusion, bottlenecks, hesitant to take action, and missing opportunities. Why does something so essential, universally valued, and deeply studied remain so elusive?

---

**CORE CONCEPT:** At the heart of this puzzle lies what behavioral scientists call the collaboration paradox: the more we push for teamwork, the more likely we are to encounter friction, miscommunication, and even declines in productivity. While our brains crave connection and collective achievement, they are also wired for self-preservation, individual recognition, and avoidance of social risk.

---

This duality means that within every team meeting, a delicate dance is taking place in the minds of everyone involved in a change process or working within a team, between wanting to belong and fearing rejection, seeking innovation while clinging to the familiar, and aiming for shared success while quietly protecting personal interests.

The paradox reveals itself in subtle and significant ways. Teams with the best intentions often become bogged down in endless coordination, overwhelmed by too many voices or tools, or paralyzed by endless debate, driven by an underlying fear of saying the wrong thing or making the wrong decision in front of their peers. Leaders may proclaim the importance of collaboration, but their actions often fail to align with their words. They hoard information, fail to take the time to clarify and answer questions, and avoid inviting cross-functional engagement. This leads to team behaviors that mirror the leader's behaviors by building walls instead of bridges.

This tension isn't a flaw to be eliminated; it's a reality to be managed. High-performing teams don't ignore this paradox; they embrace it. They bring it to the surface. They call it. Name it. Honor it. They practice with the intention to learn to balance unity with diversity, structure with flexibility, and individual ambition with collective purpose. They seek environments where it's encouraged and safe to take risks, make decisions, think things out, challenge ideas, and build on each other's strengths. Skilled team leaders and their team members know that true collaboration isn't about suppressing differences but harnessing them for greater impact and transforming them into their greatest competitive advantage.

## The Neuroscience of Performance from Relationship Bonding

During my interviews for this book, I repeatedly encountered striking parallels between sports huddles and building high-performing teams in the business world. The power of physical proximity, as vividly demonstrated through sports-style huddles, reveals how we can align with our brain's natural tendencies rather than fight against them. When teams gather in close proximity, they stimulate the release of oxytocin, often referred to as the "trust and bonding molecule," as well as Endorphins and Dopamine, the "pleasure hormones."

## NEUROSCIENCE OF TEAM DYNAMICS
### BIOLOGICAL DRIVERS OF EFFECTIVE COLLABORATION

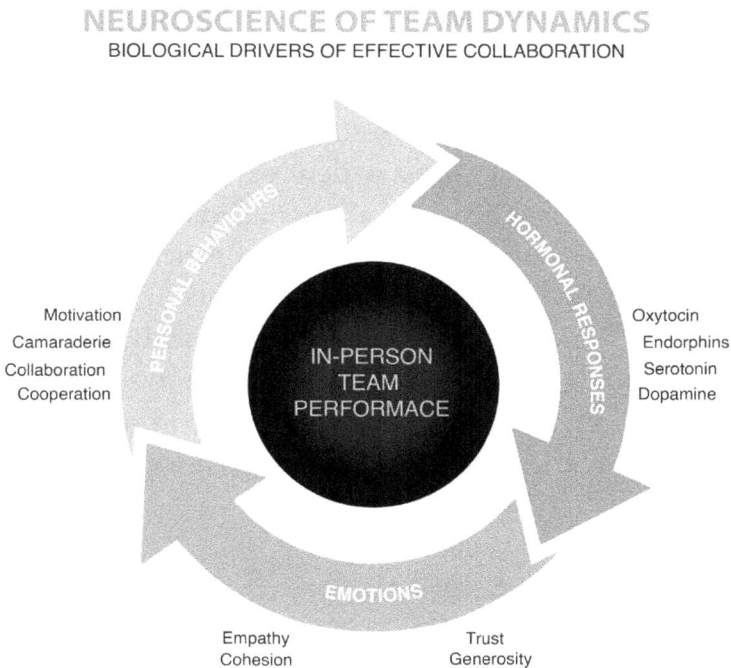

PERSONAL BEHAVIOURS

HORMONAL RESPONSES

EMOTIONS

Motivation
Camaraderie
Collaboration
Cooperation

IN-PERSON TEAM PERFORMACE

Oxytocin
Endorphins
Serotonin
Dopamine

Empathy
Cohesion

Trust
Generosity

This compelling evidence underscores the transformative impact of physical closeness on shaping team interactions. It supports why return-to-office mandates are so important to CEOs. Nothing can fully replace the wide range of positive outcomes that result from work done in close proximity with others. It is how we have evolved! This is why successful organizations, such as the Mayo Clinic, have achieved dramatic improvements in team performance by adopting neural-aware practices, resulting in a 45% increase in team innovation and an 89% boost in staff engagement.

## The Mayo Clinic Care Story

The Mayo Clinic has a large motivation for building collaborative, great problem-solving teams. The clarity of communication regarding a patient's symptoms, test results, and open discussion between various medical specialists plays a significant role in the care and healing of patients, particularly in cases that are highly complex and unique. The mission of the Mayo Clinic, "To inspire hope and contribute to health and well-being by providing the best care to every patient through integrated clinical practice, education, and research," necessitates clinical, nursing, and research teams to rethink their approach to patient care delivery.[14]

### The Mayo Clinic's Solution Approach Was Moved To:[15]

- Cross-discipline morning huddles
- Shared decision-making tools

[14]   Lin, Shih Ping, et al. "The Effectiveness of Multidisciplinary Team Huddles in Healthcare Hospital-Based Setting." Journal of Multidisciplinary Healthcare, vol. 15, no. 15, 6 Oct. 2022, pp. 2241–2247, www.ncbi.nlm.nih.gov/pmc/articles/PMC9549805/, https://doi.org/10.2147/JMDH.S384554.

[15]   Mayo Clinic Center for Innovation. (2013). Community Health Transformation: Optimized Care Team (MC6295-134). Retrieved from https://mcforms.mayo.edu/mc6200-mc6299/mc6295-134.pdf

- Patient-centered team pods
- Real-time learning systems

**The Outcomes Were:**

- Patient satisfaction increased
- Treatment innovation rate doubled
- Care costs lowered
- Staff burnout decreased

In the healthcare industry, improvements in communications and care solutions lead to a rewarding outcome—saving a life.
[16]

Now, let's examine the step of huddle-ups closely. We begin to see how this time-tested practice rapidly improves teamwork by leveraging the principles of neuroscience, specifically the concept of proximity.

## The Power of Huddle Ups

Quick huddle-ups after every play are a common practice in women's and girls' volleyball. These huddles give coaches and players a valuable opportunity to clarify team communication during challenging moments in the game. They allow the team to celebrate well-executed plays and, most importantly, offer immediate encouragement to help players move past mistakes.

---

[16] Branda, Megan E., et al. "Optimizing Huddle Engagement through Leadership and Problem-Solving within Primary Care: A Study Protocol for a Cluster Randomized Trial." Trials, vol. 19, no. 1, 4 Oct. 2018, www.ncbi.nlm.nih.gov/pmc/articles/PMC6172734/#:~:text=An%20important%20factor%20in%20implementing, https://doi.org/10.1186/s13063-018-2847-5.

This aspect of huddles is essential, as it provides players with the support and reassurance needed to quickly emotionally recover from a bad play and stay focused on the goal.

Huddles also serve as an effective moment to clarify instructions or adjust strategies on the fly in business. When used correctly, a huddle-up can create meaningful interactions that help teams become more cohesive and maintain positive momentum throughout the day or during the game.

One note: 99% of huddles are held standing up with people in close proximity. Is this your huddle norm?

The football huddle was first introduced by Gallaudet University quarterback Paul Hubbard.

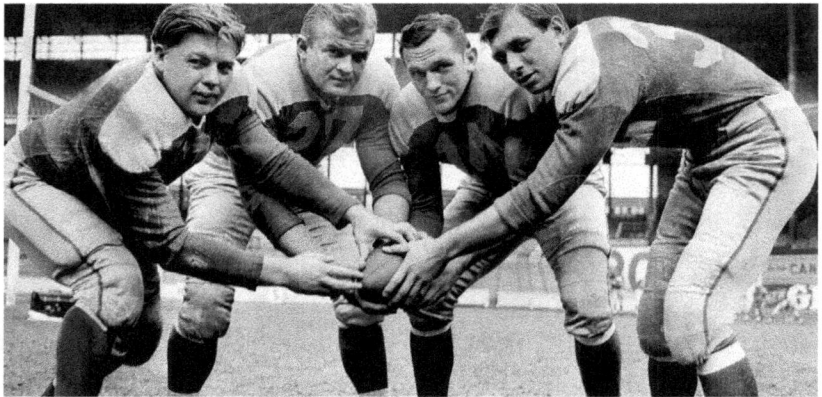

Gallaudet quarterback Paul Hubbard is credited with creating the football huddle during a season when his team faced two different deaf schools.

Studies have shown that football huddles offer several strategic and team performance benefits:

- The close proximity allows the quarterback (the leader) to ensure that all players understand their roles, enabling proper coordination for the next play.

- Direct, personal eye-to-eye contact among teammates helps build trust and camaraderie.

- During the huddle, players who are injured or struggling can be quickly identified, allowing timely adjustments and support from teammates.

- Huddles, both on and off the field, provide players with an opportunity to refocus and mentally prepare for the next play while also enabling leaders to offer motivation and sustain team energy.

Just as football teams use huddles to clarify roles, build trust, and quickly adapt, effective business huddles should be focused, brief, and inclusive—setting a clear purpose, encouraging open updates, and ending with aligned next steps—turning each gathering into a powerful tool for team connection and performance.

## Benefits of Team Members Working in Proximity to One Another

Research suggests that numerous benefits emerge when teams work in close proximity to one another. Here are just a few of the improvements you can expect by fostering an engaging in-person work culture:

- When people sit within 25 feet of high performers, their job performance increases by 15%. You can improve this number when you build a team with complementary skills and have them work in close proximity.[17]
- Face-to-face teams produce 15-20% more ideas than virtual teams, thus showing the innovative and creative power of working together in close proximity.[18]
- Companies that actively promote team collaboration are five times more likely to be high-performing organizations.[19]
- Connected teams show 23% higher profitability for the company and 66% greater well-being for thriving employees.[20]

---

[17]    Remley, Micah. "Leading Companies Prioritize In-Person Collaboration: Here's the Research." Robinpowered.com, Robin, May 2024, robinpowered.com/blog/the-science-behind-office-collaboration.

[18]    Andrews, Edmund L. "Thinking inside the Box: Why Virtual Meetings Generate Fewer Ideas." Stanford Graduate School of Business, 29 June 2022, www.gsb.stanford.edu/insights/thinking-inside-box-why-virtual-meetings-generate-fewer-ideas.

[19]    Dorronsoro, Sabrina. "The Science behind In-Person Productivity at the Office." Robinpowered.com, Robin, 30 May 2024, robinpowered.com/reports/the-science-behind-in-person-work.

[20]    Harter, Jim. "U.S. Employee Engagement Reverts back to Pre-COVID-19 Levels." Gallup.com, Gallup, 16 Oct. 2020, www.gallup.com/workplace/321965/employee-engagement-reverts-back-pre-covid-levels.aspx.

## Fight or Flight: Our Primal Brains Is Just That - Primal

Strong team leaders have the wisdom and skills of a psychologist. They understand that our brains process social threats (real or perceived) in the same way they process physical dangers. Fight or Flight. This insight enables leaders to navigate better and mitigate these seen and perceived threats more effectively.

For example, a challenging question or even a simple request for clarification ("Why are you doing this?") can trigger a neural response as a physical threat, causing strong visceral reactions, tension, and a high-pitched voice tone to be spoken. Similarly, being excluded from information or decision-making processes can activate the brain's pain centers, making individuals feel left out, unimportant, or fearful.

Armed with this knowledge, many forward-thinking team leaders support a range of team behaviors that include rapid, frequent huddles, inviting questions, addressing the unknown or unspoken, and working with, rather than opposing, our neural wiring. Strong team leaders strive to work in harmony with their team's natural brain response tendencies, creating environments and actions that foster connection, creativity, and collaboration organically.

## High Impact Collaboration Across The Silos

How do leaders leverage positive neuro behaviors and mitigate negative responses at scale across departments? Let's examine how Adobe addressed the issues that were preventing its numerous employees from collaborating effectively.

## Adobe's Kickbox Program: Democratizing Innovation at Scale

When Adobe set out to ignite breakthrough creativity across its global workforce, it faced a familiar challenge: how to empower thousands of employees to move from idea to action, not just in isolated teams but across the entire company. The solution was Kickbox—a now-famous innovation program that put tools, resources, and permission to experiment directly in the hands of every employee.

## Key Elements of the Kickbox Program

- **Universal Access:** All employees, regardless of department or role, are eligible to participate.
- **Immediate Resources:** Each participant received a red box containing a $1,000 prepaid credit card to prototype or test their idea—no management approval required.
- **Step-by-Step Framework:** The box included clear, actionable instructions for validating and pitching ideas, making the innovation process transparent and accessible.
- **Built-In Permission:** The program explicitly authorized employees to innovate, removing organizational barriers and the fear of failure.

## Impact and Results

- **2,500% Increase in Ideas Tested:** After Kickbox's launch, Adobe's capacity to validate and test new ideas surged by 25 times compared to prior internal innovation efforts.[21]
- **Patent Growth:** Adobe's global patent portfolio grew steadily, with over 9,500 patents filed by 2024, reflecting a robust culture of invention.[22]
- **Employee Retention:** Adobe consistently ranks in the top 5% of large companies for employee retention, with

[21] Skonord, Coby. "How Adobe Increased Its Ability to Validate and Test New Ideas by 2,500% - Ideawake." Ideawake, 9 Feb. 2025, ideawake.com/how-adobe-increased-its-ability-to-validate-and-test-new-ideas-by-2500/. Accessed 25 June 2025.

[22] "Adobe Patents Key Insights & Stats - Insights;Gate." Insights;Gate by GreyB, 19 May 2025, insights.greyb.com/adobe-patents/.

a recent score of 87/100, outperforming tech peers like Salesforce, Google, and Apple.[23]

- **Revenue per Employee:** Adobe's operational efficiency is reflected in its $700,000+ revenue per employee as of 2024, up from $648,000 in 2023 and $602,000 in 2022.[24]

## A Culture of Collaborative Innovation

Adobe's Kickbox program didn't just generate more ideas— it fostered a culture where employees felt empowered to collaborate, take risks, and build on each other's creativity. As one global design executive described the impact of Adobe's creative tools and innovation programs: "It helps to increase employee productivity across large teams across the enterprise."

## Conclusion

Adobe's approach to creativity and innovation is not just about tools—it's about creating systems and a culture where everyone can participate, experiment, and succeed. The Kickbox program stands as a model of how democratized innovation can drive measurable business results and lasting employee engagement.

---

[23]   "Adobe Retention Score." Comparably, www.comparably.com/companies/adobe/retention.

[24]   Bootlab. "Adobe Employee Count | ADBE." Mlq.ai, 2024, mlq.ai/stocks/ADBE/employee-count/. Accessed 25 June 2025.

# A Neuroscience Framework for Team Collaboration Success (with or without AI)

## Create Psychological Safety

This concept of developing a culture where it's safe to speak up and share ideas or ask for help can never be overemphasized. While it is easy to understand, it's hard to put it into practice and make it part of your team and organizational culture. The brain requires a stable and secure environment to function optimally. This means establishing predictable response patterns and clear expectations while providing consistent support and appreciation. Organizations that prioritize psychological safety see enhanced team performance and innovation.

## Develop Trust

Trust isn't merely a soft skill—it's a neurological necessity for peak performance. McKinsey, in its October '24 article "Go, teams: When teams get healthier, the whole organization benefits," described trust as feeling the team members can rely on one another, thus they give each other the space to get their work done and have confidence that their team member will make good judgments.[25] Yet trust doesn't happen overnight; it requires incremental steps, starting with low-risk collaboration practice exercises that gradually increase vulnerability requirements and provide proof of work delivery.

25  Smet, Aaron De, et al. "Go, Teams: When Teams Get Healthier, the Whole Organization Benefits." McKinsey & Company, 31 Oct. 2024, www.mckinsey. com/capabilities/people-and-organizational-performance/our-insights/ go-teams-when-teams-get-healthier-the-whole-organization-benefits.

## Deepen Relationship

Deepening relationships within a team goes beyond mere professional interactions; it taps into the neurobiology of connection and belonging. To foster deeper connections, effective leaders encourage open communication and the sharing of personal stories that reveal common values and experiences. This creates a common identity, which the brain associates with safety and trust. Regular team meetings, outings, hallway conversations, and even team-building activities that are both enjoyable and purpose-driven strengthen these bonds, ultimately turning colleagues into a cohesive unit. By nurturing an environment where team members feel genuinely connected, leaders can unleash a higher level of collaborative energy and commitment.

## Practice with Intentionality, and Do It Often.

The neuroscience of skill acquisition reveals that mastery of a skill requires not only practice but intentional and repetitive mind-challenging practice. Practice that is structured with clear objectives and immediate in-the-moment feedback allows our brain's neural pathways to expand over time. In the books *Culture Code* and *Talent Code*, Daniel Coyle refers to high-intensity, unconventional practice sessions as instrumental in the development of some of the world's greatest soccer players, including those from Brazil. He shared stories of how young players, who come from the poorest communities and play on very small, irregular fields, have developed highly effective, unique ball-handling skills, which he attributes to neuro-pathway formation from these unconventional practice sessions.

## Reward System Optimization

The brain's reward system, driven by dopamine, can be strategically leveraged through:

- Breaking large goals into achievable milestones
- Implementing immediate positive feedback loops
- Creating visible progress markers
- Connecting team achievements to personal growth

## Provide Neuroscience Training For Leaders

Neural-aware leaders understand brain-based team dynamics, including:

- Recognition of threat responses
- Management of emotional contagion
- Support of habit formation
- Strengthening of trust circuits

### Practical Next Steps

- Audit your current team environment for threat triggers
- Assess your trust-building mechanisms
- Map your habit-formation systems
- Evaluate your reward structures
- Measure your connection practices

## The Neural Behavior Dividends

When you build teams that work in manners that understand our natural brain responses, sometimes logical, sometimes irrational, leaders create environments where:

- Innovation flows naturally
- Change feels manageable
- Trust grows systematically
- Performance improves consistently
- Success becomes sustainable

The future of Human+AI Collaboration lies not in forcing unnatural behaviors but in designing environments, communications, tasks, and projects that incorporate coaching and mentorship, navigating the neuro-dichotomies of our brains.

## Closing Thoughts:

Before we go to the next chapter, I would like to introduce you to a podcast interview with Stephen Howe from The Wonderful Company, one of the Fortune 100 Best Workplaces to Work For®. The podcast is a perfect illustration of how applying neuroscience to collaboration and innovation can enable any team to generate important, time-saving, and money-making or saving innovations.

Stephen Howe, EVP of Human Resources and Chief Financial Officer, discusses how the company is making a positive impact on the lives of its employees and their communities through innovative programs and a commitment to sustainability.

In one standout story, field workers—those often farthest from the boardroom—devised an entirely new way to change a tractor tire by developing a specialized tire-changing truck that could come into the fields. They solved a time-consuming, costly, and persistent problem that had challenged the experienced managers. Their creative solution not only saved significant time and money but also demonstrated how frontline employees, when actually empowered and included in the problem-solving process, can develop high-value innovations from the ground up.[26]

Given the rapid changes and evolution of AI Technologies, as a thank you for your time and readership, I am providing all readers access to Transform or Die as a living, consistently updated reference resource where you will find the latest information about AI tools and platforms, practice prompts, easy-to-use scorecards, and self-directing training programs for your teams.

You can explore this chapter's topics further by delving into the guide "Navigating the Ten Neuro Leadership Dichotomies."

---

[26] Episode "The Wonderful Company's Stephen Howe on Creating a Great Place to Live and Work" originally aired January 7, 2025, as part of the "Better by Great Place To Work" podcast series.

Use this QR code or visit RussellMKern.com/transform to access the resources.

# Chapter 6

# The Unimaginable Benefits from Human+AI Collaboration

---

*"What we need to do is always lean into the future; when the world changes around you and when it changes against you... You have to lean into that and figure out what to do because complaining isn't a strategy."*

~ Jeff Bezos

---

## New Developments Every Day

In a 60 Minutes broadcast, Anderson Cooper shared a story about a neuroscience research lab in Switzerland that is helping spinal cord injury patients, who were told they would never walk, to stand and walk by implanting a neurosensor in their brain, then using AI to figure out which thoughts should trigger the correct leg muscles in the right order, bringing back the gift of mobility to severely injured patients.[27]

---

[27] Cooper, Anderson. "Injuries Left Them Paralyzed. An Early Promising Clinical Trial Is Helping Them Walk Short Distances Again." Cbsnews.com, CBS News, 11 May 2025, www.cbsnews.com/news/injuries-paralyzed-them-early-clinical-trial-helping-them-walk-again-60-minutes-transcript/.

Conversely, 60 Minutes also introduced viewers to Palmer Luckey, the 32-year-old CEO of Anduril, a defense company that builds smart, high-speed rocket drones to knock out enemy aircraft. He claims to be changing the entire concept of war by moving from "dumb" to smart, AI-empowered bombs.[28]

A decade ago, Amazon acquired a little-known Israeli chip design startup, which today underpins its AWS cloud services. Amazon is committing $100 billion—yes, billion with a B— to expand its AI infrastructure. Nvidia, the AI chip company, became the first company to reach a $4 trillion valuation. As of this writing, Apple, Meta, Microsoft, and Google are racing to scale their own AI capabilities, fueling a technology arms race that's reshaping the future of the planet.

A Wall Street Journal article titled "The Giants of Silicon Valley Are Having a Midlife Crisis of AI" reveals how the so-called Magnificent Seven of tech are scrambling—not only to harness AI's economic promise but also to defend their empires from the very disruption AI threatens to unleash.

AI is the most revolutionary, transformational utility since the creation of electricity. However, unlike electricity, which is a constant, AI is rapidly evolving, growing increasingly capable of assisting, supporting, and augmenting work.

---

[28]    Alfonsi, Sharyn, and Aliza Chasan. "Tech Billionaire Palmer Luckey Wants to Remake the U.S. Military with Autonomous Weapons." Cbsnews.com, CBS News, 18 May 2025, www.cbsnews.com/news/palmer-luckey-future-warfare-anduril-60-minutes/.

Each day, I sit down to write, and I find myself constantly feeling behind the AI curve, despite spending over 50% of my time researching and utilizing a wide range of AI tools.

Leaders would do well to heed another *WSJ* piece published that same month: Tim Higgins' "Every business leader stands at the crossroads." *The Innovator's Dilemma*, reimagined for the 21st century. Just as emerging companies once reshaped markets by bypassing legacy constraints, today's AI-driven upstarts are poised to do the same. Consider how Netflix, with its simple "movies by mail" model, quietly dismantled Blockbuster's retail empire. Disruption rarely announces itself—it arrives through a quieter, more adaptive evolution.

## The Transformations Ahead

Exactly how, and who will be the winners, with what applications and approaches, is still unknown. The chatbots of 2023–2025 were exciting and expansive, but they didn't yield results that matched the hype. In 2025, it's about agentic AI—the $12 trillion human-augmented digital workforce.[29][30]

Imagine for a minute what might become a reality as a result of AI adoption. Inventions, solutions, new levels and types

[29]   Amar, Jorge, et al. "The Future of Work Is Agentic." McKinsey & Company, 3 June 2025, www.mckinsey.com/capabilities/people-and-organizational-performance/our-insights/the-future-of-work-is-agentic.

[30]   PYMNTS. "Salesforce Says AI Is Doing 30% of Its Coding and Customer Service." PYMNTS.com, 26 June 2025, www.pymnts.com/news/artificial-intelligence/2025/salesforce-says-ai-is-doing-30percent-coding-customer-service/. Accessed 27 June 2025.

of work, never seen before. Here are a few head-spinning predictions for what to expect in the next 5 to 10 years:

- AI will enhance efficiency in everything from customer service and logistics to complex legal and financial analysis, saving up to 30% of current work hours.
- New business models like AI *"as-a-service"* for expertise (e.g., LegalGPT, MedGPT) will emerge
- AI becomes your *second brain*, proactively helping you think, decide, learn, and reflect.
- Entirely **new classes of materials**, food sources, and sustainable energy compounds are discovered.
- Entirely new industries will emerge around **"empathy engines"** that coach, console, and emotionally guide us in near real-time.
- Businesses will become **self-optimizing organisms**, where AI continually adjusts pricing, staffing, sourcing, and logistics.
- You'll be able to simulate your life paths, business decisions, and team dynamics in immersive digital twins that enable **parallel future simulations** to compare outcomes to questions like: "What if I took Job A or Job B?" and "What if we changed our culture from X to Y?"
- The educational system will shift toward **personalized lifelong learning journeys**, rather than rigid class structures and the traditional hours required to earn degrees.
- **Personalized medicine,** based on genomics and real-time AI monitoring, will become the standard of care.

- **Autonomous agents** for shopping, scheduling, customer support, and personal AI agents will manage schedules, finances, health, and even relationships, becoming increasingly common.

## The Flourishing Future: My Personal AI Prediction

Grounded in research, expert insights, and the scale of current investment, I believe AI will help humanity thrive in the coming decade. From robotic assistants to AI-driven medical breakthroughs, the potential is extraordinary. Neuroscience confirms that people flourish in collaborative environments and are naturally averse to monotonous or hazardous work, precisely the kinds of tasks AI is primed to take on. This opens the door for us to focus more fully on creativity, problem-solving, and innovation.

## What does this mean for business leaders today?

What's important for all leaders is to keep top of mind Marshall Goldsmith's world-famous CEO quote to all his CEO clients, **"what got you here won't get you there."**[31] AI is well on its way to being the most disruptive and transformative force within every business since the Industrial Revolution. It is hard for our minds to understand and comprehend the exponential impact that is forthcoming in just a few short years. The need to stop, reset your mindset, and rethink about the entire business, while difficult, is required by all business leaders now.

---

[31]   Marshall Goldsmith's quote "What got you here won't get you there": Marshall Goldsmith, What Got You Here Won't Get You There, Hyperion, 2007.

## The New Future Belongs to the Bold

While the excitement of open-source chatbots is making it easier for billions of users to access AI technologies than ever before, what isn't new is the success that comes to organizations from bold leaders who embrace a technology wave while others are still standing on the sidelines.

In the Harvard Business School case, Quigley-Simpson & Heppelwhite: The Ad Agency Model in the Age of AI,[32] Agency leaders Gerald Bagg and Renee Young were early adopters of AI to help their agency maintain its competitive advantage. Not only did it invest in AI technologies to optimize the creation and delivery of the right message at the right time in the right channel, based on deep customer insights, but it also made the bold move to hire digital natives who had an inherent understanding of the digital world.

Gerald and Rene knew their success required both cutting-edge talent and the ability to harness cutting-edge technology in collaboration. While the choices of AI tools for marketers in 2020 were exploding for a range of use cases, including prediction, content generation, automation, and investment optimization, all of which were powerful for the future growth of Quigley-Simpson's business, the options were overwhelming. The leadership had a limited budget to spend and had limited staff with the capabilities necessary to implement new AI technology. Quigley-Simpson was in the business of designing AI software.

---

[32]   HBR DAVID C. EDELMAN, JAMES BARNETT, MAY 24, 2023, Case N2-523-054

In my recent lunch meeting with Gerald, now Co-Chairman of the agency, he shared, "We were early adopters of AI technology and the hiring of digital natives. It gave us unique abilities to serve as advisors and educators, as part of being a performance agency partner to our major brand clients. Our investment in both AI tech and Digital Native talent is our secret weapon that continues to fuel our profitable growth, at a time when tech laggards in the ad industry are having to rethink their business models."

Consider the case of AES[33], a global energy company, which faced the challenge of conducting time-consuming and costly safety audits across its operations. Traditionally, these audits required significant manual effort, often taking up to 14 days to complete and incurring substantial costs. By embracing agentic AI, AES deployed an autonomous system that not only streamlined the entire safety audit process but also proactively identified compliance risks and recommended corrective actions.

The result was transformative: audit times dropped from two weeks to just one hour, costs were reduced by 99%, and accuracy increased by 10–20%.[34] This leap from traditional, reactive automation to proactive, agentic AI enabled AES to scale its operations and focus human expertise on higher-value tasks.

[33]   Regalado, Lexi. "How AES Is Transforming Utilities with AI-Driven Renewable Energy Strategies." SAS Voices, 20 Nov. 2023, blogs.sas.com/content/sascom/2023/11/20/how-aes-is-transforming-utilities-with-ai-driven-renewable-energy-strategies. Accessed 10 July 2025.

[34]   "AES Case Study." Google Cloud, 2025, cloud.google.com/customers/aes. Accessed 10 July 2025.

> **CORE CONCEPT:** The message for today's CEOs is clear—**Bold leaders who embrace not just AI but the critical importance of human-with-human-with-AI-collaboration will shape the future, while those who hesitate risk falling far behind. Winning is not about the right tech; it's about the right team dynamics that make the unimaginable a reality.**

Dag Peak, Chief Product Officer at Alianza, a cloud communication company, comments that our future with AI hinges on a balance between human judgment and intuition, with AI serving as a tool—an augmentation to human capability—not a crutch or replacement. Dag's statement reinforces the call for a shift to a blend of AI-powered, human-centered organizational cultures, where AI becomes a strategic knowledge partner and a tireless team member.

## What Humans Do Best in Collaboration with AI

AI is powerful when it comes to analyzing data, providing perspectives, and doing repetitive tasks. However, it has limitations—its capabilities are constrained by the scope and quality of its training data. AI lacks the nuanced judgment, empathy, and moral discernment that only humans possess.

The most impactful collaborations will intentionally blend these complementary strengths:

Humans contribute critical thinking, creativity, emotional intelligence, and ethical oversight. At the same time, AI has the speed to identify patterns and the capacity to process immense volumes of data, delivering outputs in ways not possible before.

Together, human and machine can form a partnership that is far greater than the sum of its parts.

## Beyond The Financial ROI of Human+AI Collaboration Investments

Organizations that invest in AI using Human+AI collaboration processes can see results quickly. Pilot programs can deliver measurable gains within 3-6 months, with full-scale transformation unfolding over 12-18 months.

Workforce Productivity increases alone from AI are now well-documented:

- Customer service agents handle **13.8% more** inquiries per hour.[35]
- Business professionals write **59% more** documents per hour.[36]
- Programmers complete **126% more** projects weekly.[37]

---

[35] Brynjolfsson, E., Li, D., & Raymond, L. (2023). "Generative AI at Work."[Brynjolfsson et al., 2023] https://academic.oup.com/qje/article/140/2/889/7990658

[36] Noy, S. & Zhang, W. (2023). "Experimental Evidence on the Productivity Effects of Generative Artificial Intelligence." https://pubmed.ncbi.nlm.nih.gov/37440646/

[37] GitHub Next (2022). "The Impact of AI on Software Developer Productivity: Evidence from GitHub Copilot." See also: "Measuring the productivity of developers with Copilot" by GitHub. https://github.blog/news-insights/research/research-quantifying-github-copilots-impact-on-developer-productivity-and-happiness/

- Consultants perform **12.2% more** tasks, **25.1% faster**, with **40% higher** quality.[38]

Even with these productivity gains, which can be viewed as financial improvements, the ROI of AI is not derived solely from its role as a productivity tool or a tool that can replace humans. It's a catalyst for the human species to code. To allow everyone on the planet to write, create, program, and easily harness nearly unlimited computing power.

The ROI of AI needs to include the results of unlocking new levels of human ingenuity and creativity, supporting designers, writers, musicians, marketers, and all kinds of knowledge workers, and developing the 'What's Next.' When leaders share the productivity benefits of AI, they gain additional returns, such as reduced burnout, enhanced job satisfaction, increased retention, and advocacy, which ultimately improve customer satisfaction and the bottom line by enabling their workforce to spend their time on inspiring and meaningful work.

**A Shared Mission for the Future**

**Using** Human+AI collaboration is a catalyst for smarter decisions, continuous innovation, and more fulfilling work for all. As of 2025, 93% of workers who use AI report that AI

---

[38]   Bommasani, R. et al. (2023). "Experimental Evidence on the Productivity Effects of Generative Artificial Intelligence." https://papers.ssrn.com/sol3/papers.cfm?abstract_id=5012601

has had a positive impact on their lives.[39] When we design systems that amplify human agency and team behaviors, we create work environments that are exciting places to be.

To support AI adoption at scale and the embrace of AI as a strategic partner within the workplace, in the pages ahead you will find the **K.E.R.N. Human+AI Collaboration Framework®**—a neuroscience-informed step-by-step process to elevate both human-with-human and human-with-AI collaboration to help solve some of your business's biggest challenges and achieve goals not thought possible.

In the next chapter, you will explore how the K.E.R.N framework and its tools align leadership goals and strategies, accelerate AI upskilling, and help transform organizational workflows for the next era.

> Given the rapid changes and evolution of AI Technologies, as a thank you for your time and readership, I am providing all readers access to Transform or Die as a living, consistently updated reference resource where you will find the latest information about AI tools and platforms, practice prompts, easy-to-use scorecards, and self-directing training programs for your teams.

---

[39] "New Global Research from Workday Reveals AI Will Ignite a Human Skills Revolution." Investor Relations | Workday, 2025, investor.workday.com/2025-01-14-New-Global-Research-from-Workday-Reveals-AI-Will-Ignite-a-Human-Skills-Revolution.

For further reading for this chapter, be sure to check out the database of case studies and research on Human+AI ROI.

Use this QR code or visit RussellMKern.com/transform to access the resources.

# Chapter 7

# Introducing the K.E.R.N. Human+AI Collaboration Framework®

The challenge for senior leaders is no longer whether to adopt AI, but how to implement it strategically, validate pilot projects, and scale solutions across the organization. Because the speed and complexity of AI-driven change defy our intuitive understanding, many leaders struggle to grasp its true transformative potential. How quickly and effectively organizations adopt and leverage AI will separate the pioneers from the passive, reshaping entire industries and fundamentally altering organizational culture, products, marketing, sales, operations, and talent strategies.

February 2025 Harris Poll of 500 Global CEOs with more than $500M in annual revenue and 500 employees reported:

- 74% of CEOs feel they will be out of a job in two years if they don't deliver measurable AI-driven business gains.

- 54% report that at least one competitor has already deployed a superior AI strategy.
- 94% of CEOs agree that AI could provide equal or better counsel on a critical decision than a human board member.

The time for Human-AI Collaboration in the flow of work is now. Having an easy-to-follow framework that addresses both the human and technical perspectives has not been available until now.

This AI-driven transformation differs from past digital transformations. It extends far beyond simply integrating AI tools. Every facet of the enterprise—from product development and service delivery to leadership, strategic planning, and customer experience—will be touched.

Success in this new era goes beyond buying the technology. It requires leaders who foster a culture of learning, curiosity, creativity, and innovation; who are willing to rethink strategies, reimagine solutions, and redesign workflows and systems to unlock AI's full potential. Cultivating a collaborative environment that encourages experimentation is key to redefining what's possible.

Given the rapid evolution of AI, this moment demands bold, innovative, and visionary leaders—leaders who go beyond merely adding AI to their technology portfolios. The challenge is to fully embrace and explore the possibilities that emerge at the intersection of human intelligence and AI's supercomputing power.

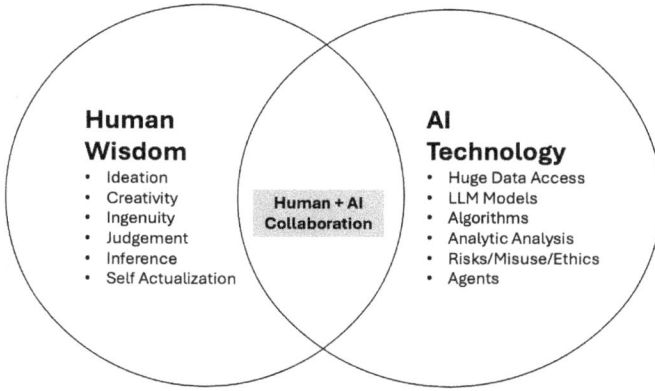

It is at the juncture between human wisdom and AI power that collaborations create new solutions, products, and answers to your most pressing challenges.

Remarkable outcomes will result from the unique teamwork that blends human creativity and AI capabilities to do together what neither could accomplish alone.

## The K.E.R.N. Human+AI Collaboration Framework®

While traditional technology adoption focuses on training, testing, and user acceptance, widespread AI adoption requires different strategies, because AI isn't just a tool to execute tasks. AI has the ability to collaborate, in essence, to co-create with its users. It can fully automate business processes, changing human roles and transforming organizational structures. AI enables users to challenge the status quo, think outside the box, and gain perspectives that have never been considered before.

Imagine the value of having your own council of the wisest men and women in the world related to your business at your fingertips to help guide you through your toughest challenges.

To unlock AI's transformative potential within the workplace requires access, fluency, exploration, and guidance from more experienced peers, and that's just for the mastery of the technology.

It also requires a structured framework for effective collaboration. Human+AI collaboration requires reinforcing the fundamentals of communications, active listening, critical thinking, high team trust, and creative ideation. In addition, it also requires upskilling in topics such as strategic thinking, workflow mapping, process improvement identification, and the ability to overcome our natural fears and resistances to change. I'd like to say that adopting and harnessing AI is simple. Unfortunately, it's complex. It takes intentionality, focused energy, and inspiring leadership to come out ahead, but the rewards are huge. Enter the **K.E.R.N. Human+AI Collaboration Framework**®: a step-by-step process designed to help CEOs and their leaders evolve and transform their organization, working in partnership with AI. The K.E.R.N. Human+AI Framework focuses on:

- The **human dynamics** within teams and across silos.

- The **expansive opportunities** that emerge when humans and AI work together based on each other's strengths and weaknesses.

Each of K.E.R.N.'s four foundational pillars is explored through two vantage points:

- The **Human Perspective** focuses on learning, curiosity, behaviors, passion, and connectedness.

- The **Technical Perspective** focuses on the foundational principles, concepts, limitations, methods, requirements, and risks.

Within each pillar concept, a set of learning tools, including scorecards, assessments, questions, AI prompts, and practical exercises, makes the K.E.R.N. framework a robust, scalable playbook for rolling out high-impact Human+AI collaboration across your organization.

**The K.E.R.N. Human+AI Collaboration Framework®
Overview**

The **K.E.R.N. Human+AI Collaboration Framework®** isn't a rigid, linear process or a simple operating system—it's a dynamic, flexible framework built around four interconnected behavioral pillars to cultivate a culture of high-impact collaboration. From my experience and research, I have found that when the actions of *Knowledge* and *Empowerment* are supported by *Reflection* and *Nurturing,* each pillar propels the next in a virtuous cycle of continuous success, growth, and innovation.

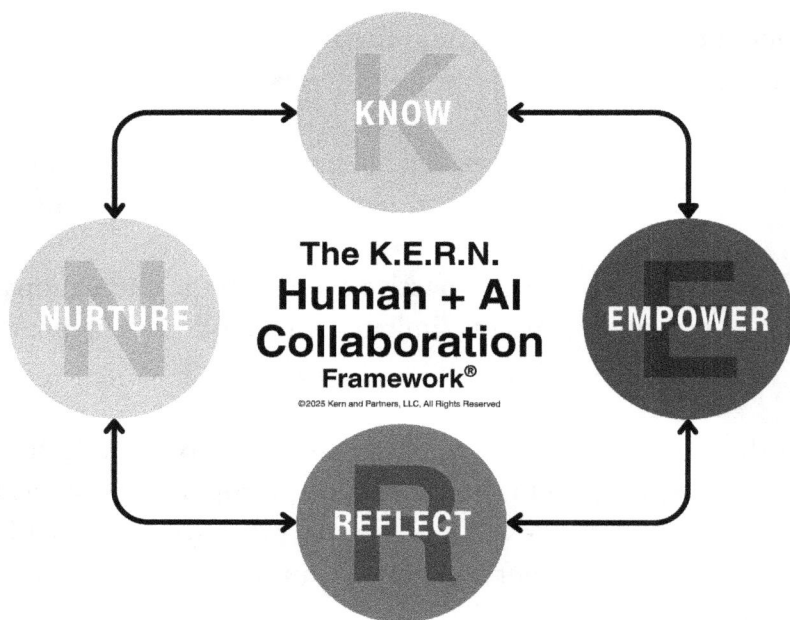

The K.E.R.N.
**Human + AI
Collaboration**
Framework®

©2025 Kern and Partners, LLC. All Rights Reserved

A business's competitive advantage does not result from simply acquiring the latest tools, but from the ability of the organization's leaders to build agile, resilient, trust-based teams. High-performance teams, where everyone is rowing in harmony.

The K.E.R.N. framework supports this development by adding AI as a strategic team member rather than just a tool in the tech stack. Each pillar plays a vital role in fostering a Human+AI collaborative team mindset:

- **Know** establishes a shared language and foundation.

- **Empower** seeks a safe space for experimentation, growth, and adaptation.

- **Reflect** promotes wisdom from continuous learning and thoughtful iteration.

- **Nurture** embeds the behaviors of appreciation and honor into culture.

Together, these four pillars shift the adoption of AI from just adding new technical capabilities to supporting the human capacity to change, adopt, and ultimately discover more effective and efficient ways to drive lasting value for customers, employees, and stakeholders.

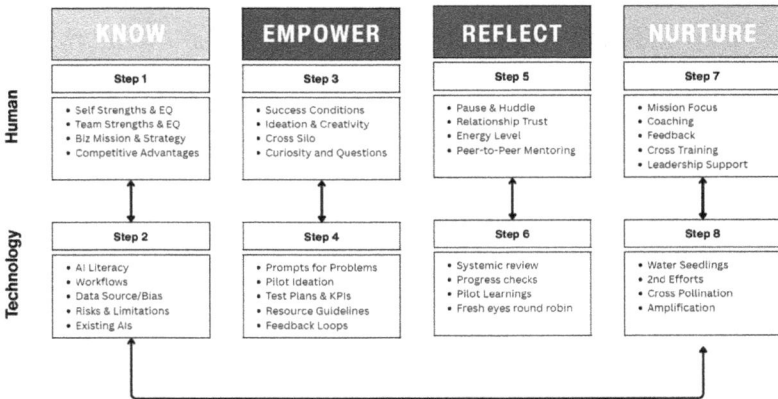

| | KNOW | EMPOWER | REFLECT | NURTURE |
|---|---|---|---|---|
| **Human** | **Step 1**<br>• Self Strengths & EQ<br>• Team Strengths & EQ<br>• Biz Mission & Strategy<br>• Competitive Advantages | **Step 3**<br>• Success Conditions<br>• Ideation & Creativity<br>• Cross Silo<br>• Curiosity and Questions | **Step 5**<br>• Pause & Huddle<br>• Relationship Trust<br>• Energy Level<br>• Peer-to-Peer Mentoring | **Step 7**<br>• Mission Focus<br>• Coaching<br>• Feedback<br>• Cross Training<br>• Leadership Support |
| **Technology** | **Step 2**<br>• AI Literacy<br>• Workflows<br>• Data Source/Bias<br>• Risks & Limitations<br>• Existing AIs | **Step 4**<br>• Prompts for Problems<br>• Pilot Ideation<br>• Test Plans & KPIs<br>• Resource Guidelines<br>• Feedback Loops | **Step 6**<br>• Systemic review<br>• Progress checks<br>• Pilot Learnings<br>• Fresh eyes round robin | **Step 8**<br>• Water Seedlings<br>• 2nd Efforts<br>• Cross Pollination<br>• Amplification |

The K.E.R.N. Human+AI Collaboration Framework® provides a comprehensive structure, steps, and roadmap to confidently integrate AI effectively into the fabric of modern organizations.

# The K.E.R.N. Collaboration Framework Pillars

## K: Know

---

*"To have harmony on a team, you need a coach who can get inside the head of every player and get them all pulling in one direction."*
— Jerry West, NBA Player

---

*"Know,"* the first pillar of the **K.E.R.N. Human+AI Collaboration Framework®**, emphasizes a key principle: high-performing teams are built on strong relationships and a deep, shared understanding. Team members know each other's mindsets, aspirations, and motivations, allowing them to anticipate actions and responses. Empathy and trust, grounded in shared experiences, are the foundation of every successful team—and are equally essential for Human+AI collaboration.

High-impact collaboration requires more than individual self-awareness and strong relationships. Teams also require a comprehensive understanding of the organization's vision, mission, purpose, offerings, history, and customer base. Without this context, it's difficult to generate new ideas or move forward effectively.

Equally important is basic AI technical fluency. This includes understanding key concepts such as machine learning, GPTs, large language models, and Agentic AI, along with the fundamentals of data structures, coding, and issues like data privacy and security.

Again, I wish I could say AI is easy. It's not hard, just a little complex.

Not everyone needs to become an AI expert. However, when team members develop a shared language around AI topics, collaboration improves. This common ground enables cross-functional teams to communicate effectively, encouraging continuous learning and ultimately leading to better results.

**Speaking the Same Language: A Real-World Example**

To illustrate the power of the **"Know"** pillar, consider this unsolicited email I received in March 2025. Although promotional in nature, it clearly captures why shared knowledge—a common language—is fundamental to effective Human+AI collaboration:

**Subject:** ✖ *The Biggest AI Bottleneck? Teams That Don't Speak the Same Language*

**Email excerpt:**

"AI initiatives don't fail because of the technology—they fail because business and technical teams struggle to collaborate. When data scientists, analysts, and business users operate in silos, projects slow down, insights get lost, and momentum stalls. Companies like OVH (40% faster AI project completion) and Aviva (5x efficiency gains) have already transformed how they work with Dataiku. Dataiku enables coders and non-coders to contribute in a unified AI workspace."

While I'm not a user of Dataiku® software, the message rings true: **the greatest barriers to success in AI are not technological—they're human.** A lack of shared understanding between technical and non-technical stakeholders is one of the most common (and costly) pitfalls.

A third critical area blends human and technology perspectives: understanding your workflows. Mapping current processes helps identify where repetitive or lower-value human tasks could be supported or replaced by AI. Engaging an expert in this mapping process is worthwhile—it uncovers high-value opportunities for Human+AI collaboration and helps teams identify impactful business cases.

Workflow analysis reduces the hype around implementation by keeping the focus on mission, measurement, and the purpose behind change.

In summary, every element of effective Human+AI collaboration—workflow mapping, technical fluency, building relationships and trust, understanding yourself as a leader, and knowing your team—requires intentional, collaborative effort. It begins with Human with Human, and then extends to Human with AI.

### E: Empower

The second pillar of the **K.E.R.N. Human+AI Collaboration Framework®**, *"Empower,"* centers on providing an environment where individuals and their teams feel confident and encouraged to explore AI's potential through experimentation.

Empowerment enables the building of the bridge between adding a technology and applying it to transform. Simply adopting AI tools and giving everyone in your organization access to AI is not enough by itself to build, test, and drive business transformations rapidly, at scale. This level of transformation requires Senior-Leader Level Vision and Support. It requires leaders to model various Human+AI collaborations that engage employees at multiple levels across the workforce. Without leaders modeling and then celebrating the wins from Empowerment, AI initiatives generally remain limited to marginal efficiency gains and fail to scale meaningfully.

Drawing inspiration from leadership principles taught by Commander Michael Abrashoff, who famously turned around the performance of a U.S. Navy ship by listening deeply and empowering his crew, this pillar emphasizes the importance of valuing team input, fostering ownership, and cultivating a sense of shared responsibility to take action. When teams feel seen and heard, they're far more likely to experiment, contribute, and innovate in partnership with AI.

From a technical perspective, **Empower** means building a culture of curiosity to find, reveal, explore, and prepare. This includes iterative refinement and the courage to test and adapt. It embraces small-scale failures as learning opportunities and promotes an "explorer mindset" across the organization.

## R: Reflect

The third pillar of the **K.E.R.N. Human+AI Collaboration Framework®** is *"Reflect"*—an intentional time out, a call to pause, to assess, and learn. Reflection is a uniquely human capacity, one that deepens understanding, strengthens relationships, and enhances collaboration. In the ever-changing and uncertain world, the discipline of reflection becomes a powerful counterbalance to reactivity.

From a **human perspective**, reflection requires teams to actually stop. Then use the time to look around, think, and honestly evaluate how they're working together as a team. Reflection is best when it's a scheduled, non-negotiable time. A reflective period allows teams to review outcomes, identify lessons learned, make necessary adjustments, celebrate progress, address any concerns, and recalibrate as needed. Research shows that intentional reflection, characterized by consistent periods of time away from work, enhances team cohesion, resilience, and trust.[40] The forthcoming chapter for this pillar provides practical steps to guide effective reflection sessions that help teams foster greater psychological safety and alignment.

From a **technical perspective**, reflection involves analyzing the progress, process, obstacles, and solutions achieved on the path within a prototype or during scaling. What's working, what's not, and why. One of the core principles of Appreciative Inquiry is that what is closely focused on and reflected upon is

---

[40]   Hartmann, Silja. "The Power of We: The Effects of Mutuality and Team Reflexivity on Team Resilience in the Workplace." Academy of Management Proceedings, vol. 2018, no. 1, Aug. 2018, p. 11708, https://doi.org/10.5465/ambpp.2018.59.

what you get more of, improve, and replicate. Simply put, if you study failures, you will experience more failures. If you study successes, progress, and positive results, you get more of that.

## N: Nurture

The final pillar of the **K.E.R.N. Human+AI Collaboration Framework**® is *"Nurture"*—the commitment to sustained growth. It's not enough to launch AI initiatives with enthusiasm; true value emerges when organizations create systems that support iterative development, enhancing both human capabilities and the impact of AI.

From **the human perspective**, nurture addresses a common challenge: initiative fatigue. Many AI projects begin with excitement, only to lose momentum within 90 days. But AI is not a one-time upgrade—it's a marathon, not a sprint. To realize its potential, leaders must consistently foster and champion conditions that enable Human+AI collaboration to remain vibrant and evolving. Neglecting this leads to stagnation, not just in technology, but in organizational relevance. History offers stark lessons: companies like Kodak, Blockbuster, and BlackBerry fell behind not for lack of innovation, but for failing to nurture those who saw the future differently and understood how consumer behaviors would change.

From **a technological perspective,** AI requires ongoing care, refinement, and tuning. A "set it and forget it" mindset does not work with AI. Instead, it will lead to decay as business conditions, customer needs, and data environments shift.

Drawing on insights from **IBM's AI Center of Excellence**[41], the Nurture pillar highlights key practices:

- Continuous feedback on data quality
- Managing model drift and bias
- Thoughtful scaling and integration
- Cross-functional alignment
- Leadership rotation to keep perspectives fresh
- Deliberate pacing of adoption to ensure long-term success

Leaders in this context are like skilled gardeners—constantly monitoring, pruning, and nourishing growth to ensure sustainability. They understand that both people and systems need attention, support, and alignment to flourish over time.

**The Journey Ahead**

The road ahead is for those organizations that embrace and understand why success will come to those that invest at the intersection of human and AI collaboration. Both forms of intelligence (human and machine) are needed to identify and propose significant use cases that turn into lasting competitive advantages.

The **K.E.R.N. Human+AI Collaboration Framework**® is designed to move the entire organization beyond isolated AI pilots that end up in the "we gave it a try graveyard."

---

[41]   IBM. "AI Center of Excellence." Ibm.com, 11 Dec. 2024, www.ibm.com/think/topics/ai-center-of-excellence.

Unfortunately, AI is complex, requires technologists, data, and security experts working in "collaboration" with strategic innovation experts. It also requires a commitment to cycles of continuous improvement, innovation, and strategic value creation.

By using the four pillars of Know, Empower, Reflect, and Nurture as a development guide mnemonic, leaders can begin (or continue) to harness AI in ways that benefit everyone the organization serves.

Given the rapid changes and evolution of AI Technologies, as a thank you for your time and readership, I am providing all readers access to Transform or Die as a living, consistently updated reference resource where you will find the latest information about AI tools and platforms, practice prompts, easy-to-use scorecards, and self-directing training programs for your teams.

For further reading for this chapter, be sure to check out the short story on the Air Force Thunderbirds' team training process.

Use this QR code or visit RussellMKern.com/transform to access the resources.

# Chapter 8

# The First Pillar – Know

*"The strength of the team is each individual member.*
*The strength of each member is the team."*
~ Phil Jackson

"Shopify Says No New Hires Unless AI Can't Do the Job:
Employees now required to integrate artificial intelligence
into teamwork, Shopify chief says."[42]

Effective Human+AI collaboration starts with a foundation of deep understanding of yourself, your teammates, your business, and the rapidly evolving technology that shapes your work.

AI is now as common in the workplace as smartphones or cloud software. Simply knowing what AI is, or that it's important, isn't enough. Leaders must move well beyond

[42] Lukpat, Alyssa. "Shopify Says No New Hires Unless AI Can't Do the Job." WSJ, The Wall Street Journal, 7 Apr. 2025, www.wsj.com/tech/ai/shopify-says-no-new-hires-unless-ai-cant-do-the-job-81c34f1e?reflink=desktopwebshare_permalink. Accessed 27 June 2025.

a passing acquaintance; they need to foster a culture where creative, confident engagement working in partnership with AI becomes the norm.

This shift is no longer a distant idea, but an urgent reality. Companies like Shopify now require employees to weave AI into their daily routines and to prove why human effort, rather than automation, should be used for certain tasks. Moves like these show how mastering AI literacy, adapting quickly, and being unafraid to experiment are now foundational to business success.

For transformative collaborations, it's not just about mastering technology; it's about knowing your business, understanding your people, and being aware of how you present yourself as a leader. True readiness in the AI age begins not with the technology, but with the mindset, curiosity, and self-knowledge of those who use it.

## The Foundation of Human+AI Collaboration: KNOW

The first pillar of the K.E.R.N. Human+AI Collaboration Framework is *Know*. At its heart, this is about building a strong foundation—understanding the core principles, essential concepts, and real-world realities that meaningful Human+AI collaboration depends on.

This pillar demands a great deal from you and your team members, both as leaders and participants in this rapidly evolving technological world. It begins with self-awareness:

understanding yourself, your team, and your business on a deeper level than just job titles or organizational charts.

It asks the same of every team member as well. It also means understanding your organization's various workflows—how work actually gets done, where human time is spent, and how processes are connected. And, crucially, it includes a working grasp of the basics of AI: what it can and can't do, key terms, and the ways it will interact with your business.

Despite what some may promise, implementing AI isn't easy. The learning curve can be steep, and the technology is often complex. That's why strong collaboration—across functions, roles, and skill sets—isn't just useful, it's essential.

Within the online living resource that you get access to with this book, you can encounter reflective questions and gain access to practical tools designed to deepen your self-knowledge and strengthen your team's collaborative capacity. These activities aren't meant to be skipped or merely considered. Real change comes when you put insights into action by: engaging with exercises, using scorecards to gauge your current state, and weaving these practices into your team's regular routines. The more you consciously assess and measure your behaviors, the more you can intentionally shape them and measure change.

**Knowledge is never static.** The best leaders never stop learning—they build, test, and refine what they know, side by side with their teams, always striving for greater impact together.

## Know

| Human Perspective | Technology Perspective |
|---|---|
| • Self Strengths & EQ<br>• Team Strengths & EQ<br>• Biz Mission & Strategy<br>• Competitive Advantages | • AI Literacy<br>• Workflows<br>• Data Source/Bias<br>• Risks & Limitations<br>• Existing AIs |

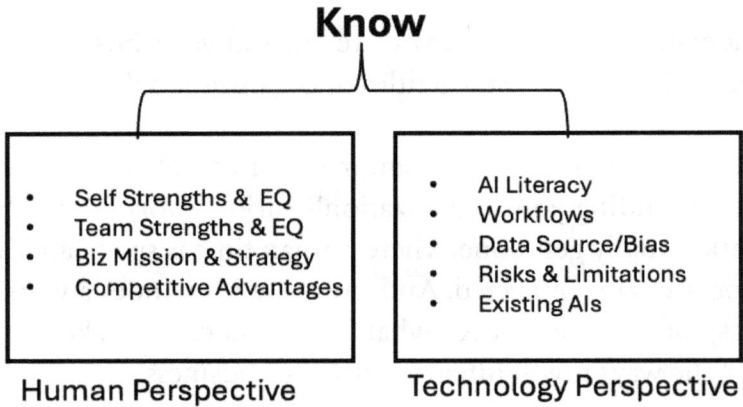

## The Human Perspective of Knowledge

### A. Know Your Business

To build a strong knowledge foundation across teams, ensure your people can answer core questions about your business's purpose, customers, value proposition, and competitive landscape. You'll find a complete set of reflective questions and tools for this chapter on the online resource page, allowing you to identify gaps and spark discussion.

Great AI implementations are not a plug-and-play solution—they require alignment across organizational, operational, and customer needs. Without a shared knowledge of the business, the problems, the bottlenecks, the culture, even the most advanced AI tools or strategies risk becoming solutions in search of a problem.

Beware of these AI Traps[43]:

- Nearly 40% of Global CEOs report AI projects delayed or cancelled due to regulatory concerns
- 60% of CEOs are very concerned about the harm to customers and employees.
- Sadly, 35% of AI initiatives are about optics (AI Washing) more than delivering meaningful business value and impact.
- CEOs consistently agree that while AI is a critical differentiator, their organizations lack the governance, planning, knowledge, skills, and oversight to execute successfully.

A Gartner® survey of over 600 organizations found that those with the deepest AI experience do not measure success by project volume or task completion. Instead, they evaluate performance through business outcomes tied to specific use cases. Further, Gartner research reveals that companies where AI teams collaborate to help define success metrics in alignment with business goals are 50% more likely to apply AI strategically.[44]

## B. Know Yourself. Know your Team Members.

Great collaborations begin with *self-awareness*. Technology can enhance human performance, but it cannot replace authentic

---

[43]   Harris Poll February 2025 Global CEO N=500. Dataiku Global AI Confession Report.

[44]   "AI Strategy for Business: 4 Key Steps." Gartner, ww.gartner.com/en/information-technology/topics/ai-strategy-for-business.

human connections, compassion, and care to build team trust, focus, and commitment to navigate increasingly complex problems.

Leaders and team members alike must regularly reflect on questions such as: Who am I as a leader or contributor? Why am I here, and how am I showing up each day? What are my values, goals, and aspirations? How do I communicate, support, and help others grow and develop? How do I give and receive feedback?

Self-awareness, emotional intelligence, and relationship awareness are the foundation of high-performing teams. When leaders invest in these areas and model vulnerability and growth, they create an environment that fosters trust, collaborative debate, and shared accountability. These are the exact behaviors that enable both human-to-human and Human+AI collaboration to thrive.

### K.E.R.N. Resource: Self-Awareness Tools and Practices

Deepening self-awareness and building strong teams don't happen by accident. It's supported by proven tools and intentional practices. For example, structured self-assessments, team mapping exercises, and communication workshops reveal hidden strengths, clarify values, and enhance trust.

Rather than crowd these pages with a long list, you'll find a curated collection of effective exercises and development tools on the book's resource site listed at the end of this chapter. By engaging with these resources, you can turn reflection into action and ensure lasting, positive change in your Human+AI collaboration skills.

> **Research Insight:** Microsoft's "Model, Coach, Care" framework, championed by CEO Satya Nadella, led to enhanced employee satisfaction by prioritizing both self-awareness and team awareness.[45] This is the type of achievement that lifts market value and productivity.

## C. Benefits of Team Members Knowing Each Other

When teams invest in self-awareness, their decision-making becomes sharper and more resilient, especially under pressure. By understanding each other's communication styles, team members create an environment of psychological safety—a proven foundation for high performance, as identified in Google's Project Aristotle research. This openness enables people to contribute freely, take calculated risks, and learn from their mistakes.

Innovation speeds up when diverse perspectives are valued and invited. McKinsey's 2023 Diversity Matters study reveals that organizations leveraging cognitive and cultural differences consistently outperform their peers in both creativity and profitability. In self-aware teams, people also learn more quickly, easily spot skill gaps, and adapt together as new technologies emerge.

---

[45]   Schwantes, Marcel. "Here's How Microsoft Knows in Less than 5 Minutes If Someone Is a Good Leader." Inc, 16 Dec. 2024, www.inc.com/marcel-schwantes/ heres-how-microsoft-knows-in-less-than-5-minutes-if-someone-is-a-good-leader/91065839.

Most importantly, engagement deepens when individual values align with the organization's purpose. This alignment fuels passion, meaning, and long-term commitment—a true advantage in navigating ongoing change.

These outcomes aren't just theory. Research has made it clear: psychological safety is essential for team effectiveness, and diversity of thought is a catalyst for innovation and improved performance. In the context of Human+AI collaboration, knowing yourself and your colleagues isn't just a personal asset—it's an essential business strategy.

> **Research Insight:** Stripe experienced growing pains during their transition from a small start-up to expanding the business. The teams struggled with tensions, which impacted their relationships and communication. Retention rates were affected. Building self-awareness and a deeper understanding of the team enabled the business to retain its top talent and achieve a customer service satisfaction score of 96%.[46]

## The Technology Perspective of Knowledge

### A. Knowledge of Technology Evolutions

Throughout history, technological revolutions have consistently reshaped how we work, create value, and communicate— from the rise of radio and photocopiers to the internet, cloud computing, and now, generative AI. Each of these breakthroughs

---

[46]   "Stripe OLT - Top Team Development | Edgecumbe." Edgecumbe, 2 Sept. 2019, www.edgecumbe.co.uk/case-study/stripe/. Accessed 27 June 2025.

has unlocked new possibilities while requiring new mindsets, workflows, and skills.

Consider this brief timeline of transformative technologies:

- **1920s**: Radio broadcasting revolutionized mass communication
- **1940s**: Programmable computers lay the foundation for modern computing
- **1970s**: Personal computers bring digital power to the individual
- **1990s**: The World Wide Web redefined global connectivity and commerce
- **2020s**: Generative AI emerges as a creative and cognitive partner in the workplace

Today, we have already entered a new frontier: an AI-enabled era marked by the rise of super agents and the approaching impact of quantum computing. The pace of change is accelerating, and with it comes both immense opportunity and growing complexity. Leaders must not only stay informed, but they must also actively guide their teams through this transformation with clarity, adaptability, and a clear vision.

## B. Knowledge of AI Technologies

The ***Know*** of technology is not intended or ask you to be an AI expert. Instead, it aims to provide you with background and perspective as a step toward AI literacy for leaders. At the core of today's most advanced AI systems are *Large Language Models* (LLMs), often featuring a ChatBot user interface, as seen in open-sourced AIs like ChatGPT-4 or Claude. These models are not sentient thinkers; they are sophisticated pattern recognition engines trained on vast volumes of text. Their primary function is to predict the next word, pixel, or phrase based on context, not to truly *understand* the information they generate.

One of the most important principles for effective Human+AI collaboration is ***inputs shape outputs***. The clarity, specificity, and structure of prompts, which are really computer code made simple, including instructions, relevant context, and examples, directly influence the quality and relevance of the AI's responses.

Equally important is selecting the right tools for the mission. Leaders should support their technical experts in evaluating various AI platforms based on how well they align with business objectives and their ability to improve existing workflows. Not all AI tools are created equal, and choosing poorly can undermine adoption and have a negative impact. Here is an abbreviated comparison table of popular open source AI tools.

| AI Model | Strengths | Best Uses | Weaknesses | Least Effective Uses |
|---|---|---|---|---|
| Microsoft Copilot (Microsoft 365) | Deep integration with Microsoft 365 (Word, Excel, Teams, Outlook), enterprise-grade security, scalability, and agent customization via Copilot Studio. Enhances productivity with workflow intelligence, reducing burnout by 19% and boosting revenue per seller by 9.4%. | Streamlining workflows in sales, HR, and operations; automating repetitive tasks (e.g., meeting summaries, data analysis); enterprise-wide productivity enhancements. | High initial and ongoing costs for smaller enterprises; occasional inaccuracies in complex tasks requiring human nuance. | Budget-constrained deployments; tasks needing deep human insight. |
| ChatGPT (GPT-4.5/o3) (OpenAI) | Exceptional versatility, advanced conversational intelligence, and robust code generation. Scales across diverse use cases with high accuracy. | General-purpose applications, creative content creation, complex problem-solving, and coding support. | Risk of hallucinations; advanced features often require premium subscriptions, limiting accessibility. | Real-time data processing; cost-sensitive operations. |
| Claude 4 (Opus/Sonnet) (Anthropic) | Superior coding capabilities, rapid response times, and strong vision processing. Excels in logical reasoning and enterprise-grade tasks. | Software development, technical problem-solving, and vision-based analytics. | Limited multimodal features; smaller ecosystem compared to Microsoft or Google. | Broad creative tasks; multimodal applications beyond vision. |
| Gemini 2.5 Pro (Google) | Seamless multimodal processing (text, images, audio) and integration with Google Workspace. Fast and efficient for productivity-focused workflows. | Multimodal analytics, Google ecosystem workflows, and customer-facing applications. | Inconsistent accuracy in complex queries; subscription-locked features. | Emotionally nuanced interactions; |

Risk management and security cannot be overemphasized. Leaders must be aware of and supportive of critical issues such as data privacy, hallucinations (AI-generated false information), result quality, consistency, and the presence of biases embedded from the training data. Responsible use of AI requires a layer of human judgment, including verifying outputs, applying ethical scrutiny, and treating AI as a collaborator, not an oracle.

> **Shadow AI is real[47]: 94% of global CEOs suspect that employees are using GenAI tools with personal logins to ChatGPT, Claude, Gemini, and others to complete company work without notice or permission.**

## C. Knowing Where the Opportunities are Through the Workflow Engineer and Mapping

Workflow engineering (mapping and refining the way work gets done) is a powerful tool used across operations, HR, finance, IT, manufacturing, and customer service to streamline tasks, clarify roles, and improve both speed and quality. While CEOs and senior leaders aren't expected to become workflow engineers themselves, having a working understanding of these concepts is vital.

With this foundational knowledge, leaders can better partner with their experts to analyze, improve, and automate business processes. The three actions of analyzing, improving, and automating are key to realizing the true value of AI and AI

---

[47]    Cawley, Conor. "Study: Most CEOs Suspect That Employees Use AI without Approval." Tech.co, 7 Apr. 2025, tech.co/news/ceos-suspect-employees-ai-approval. Accessed 15 Aug. 2025.

Agents, driving meaningful gains in both efficiency and business outcomes.

When leaders understand their organization's workflows, teams can collaborate more effectively. They can use concrete data to build strong business cases for AI initiatives, clearly identify where time is spent, spot bottlenecks, and pinpoint low-value or repetitive tasks that are well-suited for automation. This clarity ensures AI is applied where it matters most.

Below is an example of workflow mapping to illustrate this process.

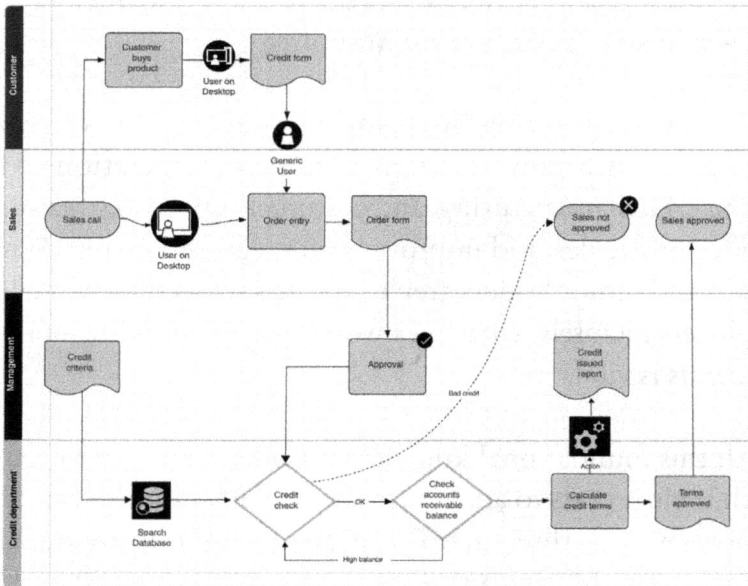

With the knowledge from workflow mapping, leaders and their teams can now start to pencil out financial AI benefits from productivity improvements, increased customer service and satisfaction, and the gains from new products, methods, services, and markets.

**A simple ROI formula to help you get started with a workflow improvement.**

Reduction in number of hours per person/week x Avg person hourly rate fully burdened x number of people who save time, times 50 weeks/year (assuming two weeks for holiday or vacation) = AI productivity dividend

Example of the financial value of small time savings per day within a medium-sized department.

5 hours of time saved/person/week x $60/hour x 25 people = $7,500 x 50 weeks = $375,500 plus

The time saved is used for higher-value work, such as strategic business improvements, innovations, and increased workforce retention and attraction.

## D. Knowing What AI Excels At

To help you identify where AI can add the most value, here are areas where AI excels:

| Area | AI Strengths |
|---|---|
| Information Processing | Rapid data organization and retrieval |
| Content Creation | Drafting, editing, and enhancing content |
| Data Analysis | Identifying trends and patterns |
| Decision Support | Scenario modeling and recommendations |
| Process Automation | Streamlining repetitive tasks |
| Customer Insights | Analyzing data for strategy and development |
| Learning & Skill Development | Personalized learning and feedback |
| Cross-functional Collaboration | Facilitating coordination and tracking progress |
| Innovation & Ideation | Generating novel concepts and approaches |
| Communication & Translation | Bridging language and discipline gaps |

Understanding these strengths allows leaders to strategically deploy AI where it can have the greatest impact.

*"To know thyself is the beginning of wisdom."*
—Socrates

## Bringing Your Knowledge Full Circle

The "K" in **K.E.R.N.** represents knowing your fundamentals. Every winning team knows and practices its fundamentals frequently, be it pitching, hitting, and catching, or shooting, dribbling, and passing.

Leaders who intentionally invest in helping their teams understand their business, themselves, their team members, and the ever-changing landscape of AI technology are the ones who lay the groundwork for transformative performance. This pillar is not about acquiring expertise in everything, but about fostering awareness, alignment, and curiosity that enable people and machines to work meaningfully together.

Next, we'll look at Empower: the second pillar of the K.E.R.N. Human+AI Collaboration Framework®.

Given the rapid changes and evolution of AI Technologies, as a thank you for your time and readership, I am providing all readers access to Transform or Die as a living, consistently updated reference resource where you will find the latest information about AI tools and platforms, practice prompts, easy-to-use scorecards, and self-directing training programs for your teams.

For further reading for this chapter, be sure to check out a curated collection of effective exercises and development tools in the KNOW section. The self-reflection questions and Knowledge Framework Assessment Scorecard will help you establish a personal or team-wide baseline. This will guide your development and ensure the next steps in your Human+AI journey are built on a solid foundation.

Use this QR code or visit RussellMKern.com/transform to access the resources.

# Chapter 9

# Second Pillar… Empower

---

*"Not finance, not a strategy. Not technology. It is teamwork that remains the ultimate competitive advantage, both because it is so powerful and rare."*
~ Patrick Lencioni, author

---

## From Worst to Best in the Navy: The Story of the USS Benfold:

In 1997, a remarkable transformation began aboard a U.S. Navy destroyer that would become a case study of the effectiveness of empowered collaboration.

Commander Michael Abrashoff had just taken command of the USS Benfold, inheriting what was widely considered one of the worst-performing ships in the Pacific Fleet. Morale was abysmally low, the safety record was concerning, and the re-enlistment rate stood at a discouraging 8%. The crew's disdain for their previous commander was so evident that during the change-of-command ceremony, they actually cheered as he departed the ship.

Twelve months later, this same ship—with the same crew—was ranked number one in performance across the entire Navy.

This extraordinary turnaround wasn't achieved through traditional command-and-control leadership, but rather through what Commander Abrashoff called "replacing command and control with commitment and cohesion."[48]

His approach exemplifies the essence of what I call *Empowered Collaboration* within the K.E.R.N. Human+AI Collaboration Framework. AI-augmented teams and teams just beginning their collaboration transformation will find the lessons in this book useful.

## Seeing the Ship Through the Eyes of the Crew

When Abrashoff first took command, he made a critical decision that would set the foundation for the behavior change and results that followed.

Instead of imposing his vision from above, he committed to understanding the ship through the perspective of those who knew it best—the sailors themselves.

---

[48]   Abrashoff, Mike. "Leaders25 Summit." Leaders25 Summit, 2025, leaders25.com/mike-abrashoff. Accessed 29 June 2025.

"I began asking each sailor three questions in my interviews," Abrashoff recounted.[49]

1. What do you like most about the USS Benfold?
2. What do you like least?
3. What one thing would you change if you could?

This wasn't just a perfunctory exercise. Abrashoff systematically interviewed all 310 crew members, taking the time to learn about their backgrounds, ambitions, and ideas.

His discoveries surprised him: "What I found out from those interviews was how smart my crew was," he noted after meeting a sailor who had scored nearly perfect on her SAT. "As soon as I realized they had the talent but lacked engagement, I understood what I needed to do."

Through this process, Abrashoff uncovered a fundamental truth that applies equally to all teams today: the knowledge, expertise, and innovative thinking needed for transformation often already exists within the organization. As the pillar Empower teaches, the challenge lies not in finding new talent or buying new tools but in creating an environment where existing talent feels empowered, encouraged, and excited to contribute.

---

[49]   D'Amelio, Tony. "HOW MIKE ABRASHOFF MADE the USS BENFOLD the BEST DAMN SHIP in the NAVY." Damelionetwork.com, 2016, blog.damelionetwork.com/how-mike-abrashoff-made-the-uss-benfold-the-best-damn-ship-in-the-navy.

## Empowering Through Invitation and Challenge

With this insight, Abrashoff established a new operating principle on the Benfold: **"It's your ship."**

These three words provided a simple-to-remember behavioral expectation that profoundly shifts accountability and ownership. In one of his first meetings with the crew, he made his expectations clear:

> "I don't care what your rank is or how long you've been on the ship—if you have an idea how we can do something better, speak up. You can come to work every day and challenge every process, every procedure, every tradition, and every custom."

The only constraint he placed on their ideas from the crew was that they couldn't exceed the budget, as it was already set. What happened next was remarkable. Crew members who had previously felt like cogs in a machine began approaching their commander with innovative solutions to long-standing problems.

Rather than providing the answers himself, Abrashoff would respond to issues with a simple but powerful question: **"What would YOU do? It's YOUR ship."** This approach encouraged his sailors to think critically and take ownership of both problems and solutions.

## From Permission to Experimentation

Permission to speak up was only the beginning. What distinguished Abrashoff's leadership was his encouragement for crew members to experiment with their ideas. He created space for them to test new approaches, learn from failures, and iterate toward better solutions, without retribution or agonizing over failure or setbacks.

One striking example involved the ship's food service.

Naval vessels weren't known for culinary excellence, but Abrashoff recognized that food quality significantly impacted crew morale and performance. Instead of accepting the status quo, he spoke with everyone involved in food preparation to understand the challenges they faced. He then gave the cooks the freedom to innovate beyond standard Navy protocols.

The Navy typically required ships to obtain bids from bulk food suppliers and accept the lowest bidder. This procurement process virtually guaranteed mediocre meals. Instead of following this procedure blindly, Abrashoff empowered his culinary team to source higher-quality ingredients within their budget constraints. The result? The USS Benfold became "a lunchtime mecca" for sailors from other ships stationed at their home port of San Diego.[50]

Another example involved an unexpected solution to a persistent maintenance problem. The ship's metal bolts would

---

[50]   D Michael Abrashoff. It's Your Ship : Management Techniques from the Best Damn Ship in the Navy. New York, Grand Central Publishing, 2012.

frequently rust, causing unsightly streaks on the paint. During one of his regular walks through the ship, Abrashoff asked a crew member for ideas to address this issue. The sailor suggested replacing the standard bolts with aluminum alternatives that wouldn't rust. Though it required a $25,000 investment, the solution extended paint jobs from weeks to nine months, saving significant maintenance time and resources.[51]

## Growing The Culture of Experimentation

These weren't isolated incidents. Abrashoff systematically supported a culture where experimentation was not just permitted but expected.

When the Pentagon imposed new requirements for arming and firing Tomahawk missiles, Benfold sailors collectively studied the training manuals and devised new methods to meet these requirements. Their approach proved so effective that it became the standard operating procedure throughout the entire Navy.[52]

During the 1997 Persian Gulf Crisis, Radioman First Class John Rafalko identified a solution to a critical communications backlog affecting thousands of operational messages. Because he was the only person who had thoroughly read the technical manuals for a new satellite system, he discovered an approach that others had missed. When he brought his idea to Abrashoff,

[51]   Smith, Sandy. "NSC 2015: Righting the Ship – from the Worst to the Best." Ehstoday.com, EHS Today, 29 Sept. 2015, www.ehstoday.com/safety-leadership/article/21917134/nsc-2015-righting-the-ship-from-the-worst-to-the-best. Accessed 29 June 2025.

[52]   Abrashoff, D Michael. "Build up Your People - Article by Mike Abrashoff from It's Your Ship." Govleaders.org, govleaders.org/build-up-your-people.htm.

the commander not only listened but also immediately contacted the chief of staff of a two-star admiral to implement the solution.[53]

## The Compounding Results of Empowerment

As explained throughout his book, *It's Your Ship*, the cultural transformation aboard the USS Benfold yielded measurable results that extended far beyond improved morale:

- Re-enlistment rates skyrocketed from 8% to nearly 100%
- The ship won the prestigious Spokane Trophy for combat readiness
- Operating costs were reduced by 25%
- The promotion rate for crew members tripled

Perhaps most tellingly, when Abrashoff departed the ship after his tour of duty, his change-of-command ceremony lasted just 45 seconds rather than the customary 90 minutes. His entire speech consisted of just five words: "You know how I feel." As he later learned from a crew member's email, "there wasn't a dry eye on the boat" as he left.

## The Essence of Empowered Collaboration

The story of the USS *Benfold* exemplifies the transformative power of *Empowered Collaboration*—what becomes possible when leaders cultivate an environment where individuals are

---

[53]   D Michael Abrashoff. It's Your Ship : Management Techniques from the Best Damn Ship in the Navy. New York, Grand Central Publishing, 2012.

not only allowed but actively encouraged to bring forward their best thinking, experiment with new approaches, and take full ownership of outcomes.

With AI now joining the crew on your business ship, the principles of empowering collaboration are just as vital today as they were in 1997. Commander Michael Abrashoff's story reminds us that organizations already possess remarkable reserves of human talent and initiative. What's needed is an environment where this potential can be unleashed. Building collaboration skills and creating opportunities for teams to practice working together are key first steps in tapping into this collective strength.

As teams begin integrating AI tools, a new level of collaboration becomes possible, harnessing both human creativity and the generative power of intelligent technology. When Human+AI Collaboration is intentionally empowered and skillfully practiced, AI doesn't just automate routine tasks; it magnifies what humans can achieve, solve, and create. By fostering trust, autonomy, and open experimentation, leaders can unlock unprecedented outcomes and set their organizations on a course for continuous innovation.

The key to *Empowered Collaboration*, as Commander Abrashoff demonstrated, lies in leadership behaviors that:

1. Genuinely seeks to understand the perspective of team members
2. Invites everyone to challenge established processes and propose alternatives

3. Creates space for listening, experimentation, and calculated risk-taking
4. Recognizes and celebrates initiative, even when experiments don't yield the expected results
5. Builds a culture where ownership is distributed and success is shared

As we navigate the rollout of AI and AI Agents into our teams and workflows, the story of the USS *Benfold* offers a timeless reminder: the most powerful asset in any organization isn't its tools or processes—it's the collective intelligence, creativity, and commitment of its people.

When individuals are truly empowered, there are few limits to what a team can achieve. Like Commander Abrashoff, today's leaders must be willing to say, *"It's your ship,"* inviting ownership, experimentation, and initiative at every level.

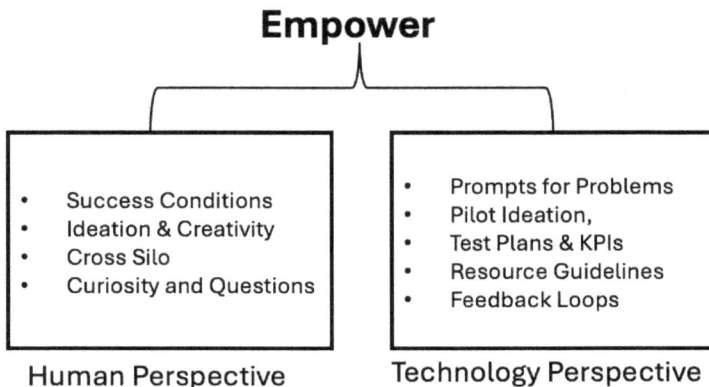

## Empower

| Human Perspective | Technology Perspective |
|---|---|
| • Success Conditions<br>• Ideation & Creativity<br>• Cross Silo<br>• Curiosity and Questions | • Prompts for Problems<br>• Pilot Ideation,<br>• Test Plans & KPIs<br>• Resource Guidelines<br>• Feedback Loops |

## Empower Your Teams to Push Boundaries

The lesson of the USS Benfold is not merely that a good leader says, "It's your ship"—but that true empowerment happens when everyone is given both the responsibility and the encouragement to question, explore, and shape new ways of working. Abrashoff didn't just hand over ownership; he invited his crew to see possibilities where others saw limits, and he stood behind them as they tried new things, even when outcomes were uncertain.

This active, empowered approach is at the heart of the K.E.R.N. Framework's Empower pillar. It's not passive permission, but an invitation to every member of the team—human or AI-enhanced—to think boldly, try without fear, and shape the future together. In today's world of Human+AI Collaboration, the Empower pillar charges leaders and teams to:

- Encourage curiosity and responsible risk-taking in partnership with AI.
- Recognize and celebrate both successful experiments and valuable lessons from what didn't work.
- Distribute decision-making so that innovation flows from everywhere, not just the top.

Just as Abrashoff's crew transformed their ship through empowerment and exploration, so must we empower our teams—human and AI alike—to continuously learn, adapt, and push boundaries. Building an environment where everyone can say, "I have the autonomy to try," is what will set apart organizations that thrive in the Human+AI era.

Empowerment isn't just a feel-good sentiment—it's the launchpad for exploration, adaptation, and real innovation. When we engage the Empower pillar, we're not just inviting ownership—we're unlocking possibility.

## Your Next Step

Before we proceed to examine the next pillar, Reflect, please answer these self-reflection questions. The answers will help guide you as you move through the K.E.R.N. Human+AI Collaboration Framework.

### Empowerment From a Human Perspective

1. **Success Conditions** - Have we clearly defined what success looks like for our team, and do all members feel empowered to contribute toward achieving it?

2. **Ideation & Creativity** - Do we consistently generate and nurture new ideas in a way that encourages originality rather than conformity?

3. **Cross-Silo Working** - How effectively do we collaborate across functions and departments, ensuring everyone has both the access and authority to contribute?

4. **Curiosity & Questions** - To what extent do we encourage team members to ask questions that challenge the status quo and deepen our understanding of the problems we face?

## Empowerment from a Technology Perspective

1. **Prompts for Problems** - How well do we use AI-assisted prompts or questions to expand our thinking when defining and diagnosing problems?

2. **Pilot Ideation** - Are team members empowered to propose and initiate AI pilots, even if the ideas feel unconventional or risky?

3. **Test Plans & KPIs** - Do we establish clear test plans and success metrics (KPIs) before starting AI experiments, so learning and impact can be measured?

4. **Resource Guidelines** - To what degree are resources (time, tools, training, and budget) made available so our team can responsibly experiment with AI?

5. **Feedback Loops** - How effectively do we capture lessons from AI pilots and share them across the team so that failures fuel learning and successes can be scaled?

Given the rapid changes and evolution of AI Technologies, as a thank you for your time and readership, I am providing all readers access to Transform or Die as a living, consistently updated reference resource where you will find the latest information about AI tools and platforms, practice prompts, easy-to-use scorecards, and self-directing training programs for your teams.

You'll find tools, checklists, and activities to help leaders cultivate the Empowered Collaboration mindset in the EMPOWER section.

Use this QR code or visit RussellMKern.com/transform to access the resources.

# Chapter 10

# Third Pillar—Reflect

---

*The Pause that Powers Human+AI Collaboration*

---

## When Reflection Is Life or Death

On a freezing morning in 1986, the Space Shuttle Challenger launched into history… and tragedy. Despite urgent warnings from engineers about the effects of the cold on critical rocket components, leaders pressed "go," silencing doubt in the name of momentum. Seventy-three seconds later, the shuttle exploded.

The disaster was not just technical; it was a catastrophe of missed reflection—a warning to every leader facing high-stakes decisions powered by rapidly advancing technologies.

In a world being shaped in ways we can't imagine by new technologies, the cost of a lack of collaboration, failing to pause, reflect, and listen with respect to each other, without our personal agendas or egos overriding the information coming from all team members, grows ever greater.

## The Power of Reflection in Human+AI Teams

*Reflect* is the third pillar in the K.E.R.N. Human+AI Collaboration Framework®, which calls for a deliberate pause. This pillar is not about nostalgia or idle review. It is the practice of intentionally stepping back to examine, learn, and adjust before charging forward.

### Why is reflection so urgent now?

Teams today must navigate rapid technological change, shifting roles, and complex human dynamics. The best teams learn from studying what works, and also from what doesn't, together, again and again. Reflection is the pause between going forward with decisions, actions, adjustments, and adaptation, which is essential for both strong relationships and positive results.

> **Reflection turns experience into insight—and insight into action.**
>
> What is a Reflection Meeting?
>
> Moments in time, where the pressure of getting it done is set aside to focus on what is right now. What is the state of the team's relationships? What changes in the relationships have transpired? Are we feeling energized by the changes? What has been accomplished from a technology perspective? Where are we today? When we zoom out, are there adjustments to our experiments, approaches, and assumptions that need to be updated?

## Why Reflect? The Research and the Stakes

Structured reflection is the key to the difference between progress and stagnation in any work process, team effort, and especially within a transformational initiative. Reflection in Human+AI teams means taking the time to look through two lenses—examining both team relationships and technology performance to learn what is working well and doing more of that, as well as what needs to change to help the team and the tech improve together, tomorrow.

Navy SEALs use After Action Reviews (AARs) to assess their missions and learn from both successes and failures. Their debriefs focus on three key questions:

**What went well?** This question helps the team identify successful strategies and tactics that can be replicated in future missions.

**What went wrong?** By examining mistakes or missteps, the team can identify areas for improvement and adjust their approaches accordingly.

**What can be done better next time?** This focuses on actionable steps for improvement, emphasizing adjustments that can enhance future performance.

~Jocko Willink and Leif Babin. Former Navy SEAL commanders

Numerous case studies and statistics show just how vital using reflection is to ensure teams operate at a high level:

- **MIT/Boston Consulting Group (2020)**[54]:
  Organizations gaining high value from AI all structure mutual learning between people and technology.

- **Stanford Behavior Design Lab (2023)**[55]:
  Teams with positive relationship habits saw 39% higher productivity and 45% faster adaptation to change.

- **Google's Project Aristotle:**
  Psychological safety—built through honest reflection—is the top predictor of team effectiveness.

- **Deloitte (2024)**[56]:
  Without pausing to reflect on both the tech and human sides, 70% of enterprise AI efforts stall at "proof-of-concept."

- **Salesforce (2023):**
  Integrating AI habit-tracking and relationship mapping improved team effectiveness by 38% and customer satisfaction by 42%. These outcomes were detailed in *Harvard Business Review's* October 2023 feature,

[54] Ransbotham, Sam, et al. "Expanding AI's Impact with Organizational Learning." MIT Sloan Management Review, 2020, sloanreview.mit.edu/projects/expanding-ais-impact-with-organizational-learning/.

[55] https://behaviordesign.stanford.edu/

[56] Deloitte. "State of Generative AI in the Enterprise 2024." Deloitte, 2024, www.deloitte.com/us/en/what-we-do/capabilities/applied-artificial-intelligence/content/state-of-generative-ai-in-enterprise.html.

which offers a comprehensive analysis of Salesforce's methodology and insights.

**What Does This Tell Us?**

New technology will not transform organizations. The difference between leaders and laggards lies in how leaders and their teams—human and machine—learn, adapt, and do better together through the intentional action of reflection.

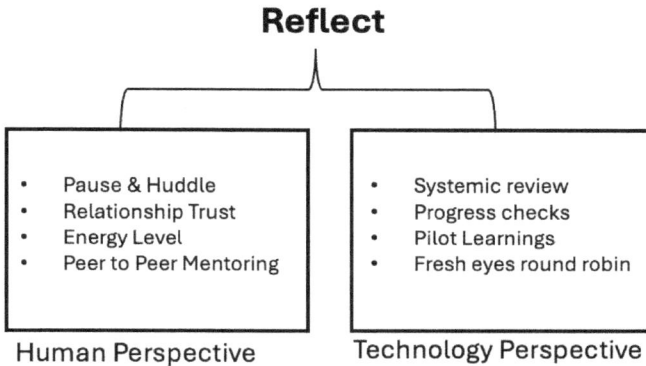

## Reflect

| Human Perspective | Technology Perspective |
|---|---|
| • Pause & Huddle<br>• Relationship Trust<br>• Energy Level<br>• Peer to Peer Mentoring | • Systemic review<br>• Progress checks<br>• Pilot Learnings<br>• Fresh eyes round robin |

# Reflection from the Human Perspective

**Why It Matters:**

Strong team relationships don't sustain themselves automatically. Trust, communication, and shared understanding must be built, especially when new technologies are transforming the landscape.

**When neglected:**

- Psychological safety erodes
- Critical information is withheld
- Risk and conflict increase
- Innovation stalls

**When valued and practiced**, reflection deepens trust and unlocks creative risk-taking—the very thing that drives innovation in the AI era.

### Key Human Reflection Activities & Practices

### Structured Relationship Retrospectives

- Appreciation Circle: Each member shares one appreciation for a teammate's contribution, reinforcing strengths.
- Relationship Radar: Team plots areas of trust/ communication to surface blind spots and growth areas.
- Impact Feedback: "When you... I feel... Because..." statements facilitate honest, blame-free exchanges.

### Relational Check-ins

- Temperature Checks: 1-5 ratings on trust, safety, or communication; discuss any drops as a group.
- Paired Walks: Informal, rotating partner conversations guided by prompts about how the team is working together.

- **Assumption Testing:** Teams safely surface and challenge unspoken beliefs about each other.

## Relationship Agreements

- <u>Review and update</u> explicit agreements on norms, feedback, and conflict. (Make expectations visible, not hidden.)
- Quarterly Agreement Reviews: Assess what's working, what's slipping, and reset as needed.

### Case Study: Relationship Reflection at Buurtzorg57

The Netherlands-based healthcare organization Buurtzorg provides a compelling example of the impact of relationship reflection. Their self-managing nurse teams (typically 10-12 nurses) serve patients with minimal management overhead, requiring extraordinary teamwork and trust.

Each Buurtzorg team dedicates monthly "relationship hours" in addition to operational meetings.

These sessions follow a simple but powerful format:

**Appreciation Round:** Starting with specific appreciations for contributions observed since the last meeting

---

[57]   Gray, Bradford, et al. "Home Care by Self-Governing Nursing Teams: The Netherlands' Buurtzorg Model | Commonwealth Fund." Www.commonwealthfund.org, 29 May 2015, www.commonwealthfund.org/publications/case-study/2015/may/home-care-self-governing-nursing-teams-netherlands-buurtzorg-model.

**Relationship Tensions:** Naming tensions in how the team is working together

**Relationship Experiments:** Designing small experiments to address tensions

**Agreement Review:** Revisiting and potentially revising team agreements

Initially skeptical about dedicating precious time to relationship discussions, teams found that these sessions significantly reduced conflict, improved information sharing, and enhanced their ability to handle challenging situations. When Buurtzorg began introducing AI tools for care planning and documentation in 2022, teams with strong relationship practices adapted far more successfully than those who had neglected this dimension.

One nurse reflected: "When the AI tools first arrived, we had fear and resistance like any team. But because we already had a practice of talking openly about our relationships and concerns, we could discuss how these new tools were affecting our team dynamics. Teams without this practice got stuck in unspoken anxieties and resistance."

Buurtzorg's experience demonstrates that relationship reflection isn't just about feeling good—it creates the foundation for navigating complex adaptive challenges, including technological transformation.

# Reflection from the Technology Perspective

**Why It Matters:**

Reflection is not only for people. Teams must also step back and discuss: How are we actually using these new tools? Are the results meaningful, or merely technical? Is everyone growing, or are we leaving some teammates behind?

**When neglected:**

- **Pilot Purgatory:** Endless pilots, few real-world results
- **Misaligned Value:** Tools succeed, but miss the organization's real goals
- **Unequal Adoption:** Gaps grow between "AI natives" and "AI outsiders"
- **Lost Learning:** Mistakes and wins aren't shared, so progress stalls

## Key Technology Reflection Activities & Practices

**Multi-Horizon Progress Reviews:**

- *Weekly "Standups":* What did we try, what did we learn, what needs work?
- *Monthly Integration Reviews:* How fully are AI tools woven into daily workflow?
- *Quarterly Value Assessments:* Are AI investments delivering business, customer, and team benefits?

## Three-Dimensional Progress Tracking:

- *Technical Implementation*: Usage metrics, system reliability
- *Capability Development*: How is AI fluency and comfort spreading?
- *Value Creation*: Are we achieving new value and not just efficiency?
- *Relationship Metrics*: How does AI affect group dynamics?

## Learning Acceleration Rituals:

- *AI Experience Exchanges:* Teams share lessons (wins and failures) with others
- *Failure Valorization:* Celebrate what was learned from what didn't work
- *Cross-functional Panels/External Benchmarking*: Broaden reflection by tapping wider experience

**Case Study: Salesforce's Relationship-First Transformation**[58]

In 2023, Salesforce launched a "relationship-first" initiative to accelerate Human+AI collaboration across its teams. They combined AI-powered habit tracking—monitoring how teams communicated, gave feedback, and adapted to change—with regular, structured

---

[58]  Salesforce. "Customer relationships are frayed. AI to the rescue?" Salesforce Blog, December 21, 2023

reflection sessions focused on both team dynamics and technology use.

The result? Teams reported a 38% boost in effectiveness and a 42% increase in customer satisfaction within months. These structured moments allowed for pauses and reflection, enabling the review of not only technical progress but also how team members connected, innovated, and supported one another, thereby strengthening trust and surfacing valuable lessons quickly.

Reflection isn't just about reviewing outcomes; it's about building deeper alignment among humans and harnessing technology to drive sustainable, measurable impact.

## Integrating Reflection for Sustainable Success

### Don't Separate Data from Dialogue

The biggest failures (like Challenger) happen when people are pressured to move fast, avoid tough questions, or separate "facts" (the data) from "feelings" (the team's lived experience).

Integrated reflection means:

- Pausing together to look at both numbers and narratives.
- Inviting every voice, not just the loudest or the most technical.
- Asking: What does the data say, and what does our team feel about it?

- Surfacing hesitant opinions, doubts, and new ideas—on both human and technical fronts.

When reflection is routine and integrated, teams build "collaboration intelligence." "Collaboration intelligence" refers to the ongoing capacity to adapt, innovate, and stay aligned, regardless of changes in tools or stakes.

## Closing Inspiration

*"To reflect is not to slow down progress; it is to unlock the momentum that matters. Rest nurtures creativity, which nurtures activity. Activity nurtures rest, which sustains creativity. Each draws from and contributes to the other."*
—Kim John Payne, Simplicity Parenting

The leaders who will thrive are not those who move the fastest, but those who know when to pause and reflect. For effective Human+AI collaboration, reflection is not a luxury—it's the superpower that protects what matters most and empowers what's next. The **Reflect** pillar leads to the **Nurture** pillar, which we're looking at next.

Given the rapid changes and evolution of AI Technologies, as a thank you for your time and readership, I am providing all readers access to Transform or Die as a living, consistently updated reference resource where you will find the latest information about AI tools and platforms, practice prompts, easy-to-use scorecards, and self-directing training programs for your teams.

You can find a Reflective Practice Toolkit for this chapter that includes:

10 Questions for Leaders on Reflection from a Human Perspective
10 Questions for Leaders on Reflection from a Technology Perspective
Team Reflection Effectiveness Assessment Scorecard

Use this QR code or visit RussellMKern.com/transform to access the resources.

# Chapter 11

# Fourth Pillar... Nurture

---

*"The best time to plant a tree was 20 years ago.*
*The second best time is now."*
~ Chinese Proverb

---

The fourth pillar within the K.E.R.N. Framework for Human+AI Collaboration (***Nurture***) asks leaders to prioritize the importance of nurturing growth, intentionally supporting both human intelligence and technological knowledge. Unlike the general 'growth mindset,' nurturing growth here means actively and continually developing people and ideas alongside advancing technologies.

Within this pillar, nurturing means creating safe spaces for experimentation, where new ideas can grow together, both for humans and AI systems. This chapter examines the concept of consciously cultivating the conditions for sustainable innovation, acknowledging that genuine growth emerges not from passive hope or isolated efforts but through deliberate care, targeted support, and adaptive leadership.

> The word "nurturer" encompasses a set of behaviors that involve encouragement and support. Collaborations between team members require nurturing to help colleagues listen, stay curious, and not rush to a solution.

Nurturing the new, whether it be ideas, pilots, or processes, is required because humans often reject something new, unfamiliar, or different. "This won't work." "No way." "That's crazy." "It's never been done." "We don't do it that way." "It's too expensive." The list of idea-killing phrases is endless; it can quickly drain the energy from even the most promising ideas.

Just as a gardener nurtures their seedlings and tends to their garden throughout the seasons, so too is the requirement for Human+AI Collaboration. Leaders need to provide the soil, water, sunlight, warmth, and climate to enable the kernels of new ideas to germinate and sprout.

> **Inspiration from Gardeners Not Mechanics.** *How to cultivate change at work by Gary Lloyd*
>
> Gardeners are quiet strategists of growth. They understand the science behind transformation, including germination, photosynthesis, and the intricate mechanics of thriving ecosystems. Yet they also embrace the art of timing, care, and responsiveness. With the right tools, they prepare the soil and create conditions for growth, balancing light, water, space, and the influence of surrounding plants. But their work doesn't stop at planting. They observe. They adjust. They respond to changing environments.

Is the weather unusually hot or dry? Are weeds competing for nutrients? Are plants flowering but failing to bear fruit? Should planting be staggered to avoid a harvest glut? Gardeners ask these questions daily, not because they're uncertain, but because they know adaptability is essential to resilience.

They follow a method, rooted in fundamentals, but responsive to nuance. They adjust for the season, the climate, and the plant. And that is precisely the mindset required for nurturing high-value collaboration in Human+AI teams.

Ideas, like seedlings, are vulnerable when new. They need consistent care, encouragement, and protection from premature criticism or neglect. A leader with a gardener's mindset knows when to stake an idea to help it stand, when to prune for clarity, and when to step back and let it flourish. It's not about forcing growth but creating the right conditions for it to emerge, take shape, and bear fruit.

This metaphor isn't just poetic; it's deeply practical. Leaders who embrace a gardener's mindset cultivate cultures where innovation can take root and thrive, even amid uncertainty and change.

The same is true for organizational transformation. Human+AI Collaboration is a continual, generative, and back-and-forth process between team members and various AI tools to create business-critical solutions to significant challenges.

## From Skepticism to Sensation - The Savannah Bananas' Journey of Embracing the Unconventional

In 2016, Savannah, Georgia found itself without a baseball team when the Savannah Sand Gnats relocated. Seizing the opportunity, Jesse and Emily Cole introduced the Savannah Bananas[59], a team that would soon revolutionize the sport.

Initially competing in the Coastal Plain League, the Bananas quickly distinguished themselves by infusing entertainment into the game. Players performed choreographed dances, and the team introduced unique fan experiences, such as the "Banana Baby" and the senior citizen dance team, the Banana Nanas.

Despite early skepticism and criticism likening them to a baseball version of the Harlem Globetrotters, the Bananas persisted. In 2022, they fully embraced their innovative approach by leaving the Coastal Plain League to focus exclusively on "Banana Ball," a fast-paced, entertainment-driven version of baseball with unique rules designed to enhance fan engagement.

This bold move paid off. The Bananas have since sold out every game, including performances in major league stadiums, and have garnered a massive following on social media platforms like TikTok. Their success highlights the value of embracing unconventional ideas and prioritizing the fan experience.

---

[59]    https://thesavannahbananas.com/about_us/

The Savannah Bananas' journey from a fledgling team to a nationwide phenomenon serves as a testament to the impact of innovation and the courage to challenge traditional norms. Their story illustrates how nurturing unconventional ideas can lead to extraordinary success.

## The Path to Nurturing Sustainable Human+AI Collaboration is Rarely Linear

According to Deloitte's research, a large majority of respondents (68%) say their organizations have only been able to move 30% or fewer of their GenAI experiments fully into production.[60]

The challenge isn't simply engaging in Human+AI collaboration—it's cultivating the outcomes of that collaboration into meaningful, differentiated, and enduring organizational capabilities. The real work begins with what comes before the pilot, where early ideas are validated through the mapping of workflows, confirmation of the data, and nurturing of constructive debate about what changes in a system, as well as what processes and technologies can grow, evolve, and deliver long-term organizational value.

This chapter focuses on how leaders can best nurture the development of not only human talent but also the

---

[60]   "Deloitte Washington National Tax Practice Names Todd Keator as Principal and Leader of Sec. 1031 Exchange Practice - Press Release | Deloitte US." Deloitte, 20 Aug. 2024, www.deloitte.com/us/en/about/press-room/state-of-generative-ai-Q3. html. Accessed 30 June 2025.

technological proficiencies needed to sustain change over time. Ultimately, *Nurture* is the foundation of continuous adaptation and evolution. Without it, even the most promising AI pilots risk becoming one-off experiments—initiatives that flash briefly, then fade—rather than becoming transformative capabilities that redefine how the organization works, learns, and leads.

## Nurturing Growth from the Human Perspective

Achieving meaningful success with Human+AI Collaboration requires a fundamental shift in leadership mindset. Rather than seeing AI adoption as a one-off, stand-alone project or simply another addition to the technology stack, leaders must embrace it as the beginning of an ongoing, co-creative partnership between people and technology. This shift means that team missions, processes, and behaviors aren't temporary or isolated. Instead, they become dynamic and evolving.

Leaders may find it challenging to adopt this mindset. Many business environments demand immediate ROI and are under constant pressure to deliver quick results. Yet research consistently shows that the greatest value from AI doesn't appear at the point of deployment. Instead, it comes through sustained effort: ongoing iteration, continuous learning, thoughtful improvements, and a focus on long-term stability. Meaningful Human+AI Collaboration is not a one-time achievement, but a journey measured by growth and adaptation over time.

*Nurture* empowers leaders to shift:

- **From Projects to Patterns**: Human+AI Collaboration is not a one-time initiative; it is a set of evolving practices that should become embedded into the organization's daily rhythms and ways of working.

- **From Control to Cultivation**: letting go of rigid implementation models in favor of a more responsive, cultivation-based approach, guiding growth, adapting to change, and stewarding progress over time.

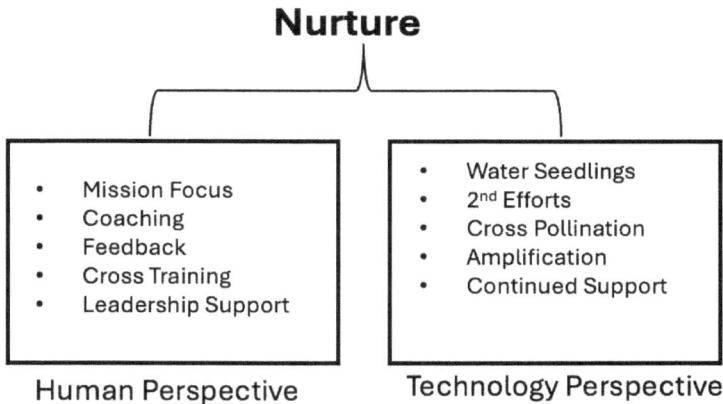

## Nurture

| Human Perspective | Technology Perspective |
|---|---|
| • Mission Focus<br>• Coaching<br>• Feedback<br>• Cross Training<br>• Leadership Support | • Water Seedlings<br>• 2nd Efforts<br>• Cross Pollination<br>• Amplification<br>• Continued Support |

## Nurturing Growth and Success: The Human Perspective

While a strong technological foundation is important, the true depth and effectiveness of Human+AI Collaboration depend on human capability. Developing these capabilities is an ongoing process, requiring thoughtful investment and sustained care.

As organizations adopt AI-driven approaches to working, job roles and responsibilities naturally evolve, necessitating new skills and growth opportunities across teams. By investing in upskilling and fostering a culture of learning, leaders can empower their people to grow confidently alongside advancing technologies, reducing apprehension, and inspiring continued engagement.

Today, leadership is less about directing and more about facilitating. The best leaders foster independence, critical thinking, trust, and constructive debate. They nurture learning ecosystems where ongoing development is woven into daily work, and everyone can experiment, share insights, and grow together.

At the heart of this growth is psychological safety and trust between team members to play their position and do their jobs with expertise. When team members feel valued and safe to voice ideas, ask questions, or share concerns, true collaboration among people and with AI can flourish. Nurturing this sense of safety is a vital foundation for successful, sustainable Human+AI Collaboration.

## Nurturing Growth and Success: The Technology Perspective

Even with clear principles and promising examples, sustaining Human+AI collaboration from a technological standpoint presents significant challenges. The complexity of implementing AI at scale, combined with the rapid pace of technological evolution, requires leaders to anticipate and actively manage several common pitfalls.

One of the most persistent challenges is the **phenomenon of initiative fatigue, characterized by its short-lived nature**. Many organizations begin their AI journey with enthusiasm, only to see interest fade as attention shifts or competing priorities emerge within 90 days. Without intentional effort to reinforce learning and celebrate progress, promising initiatives often lose momentum before they deliver meaningful value.

Another risk is the **double digital divide**—a gap that exists not only in technological access, but in the organizational readiness to use it effectively. It's not enough to have the tools; teams must also have the cultural, emotional, and strategic capabilities to apply them in context.

Many organizations also fall into the trap of **pilot purgatory**, where endless experimentation yields insights but no scalable action. Teams may launch dozens of pilots without ever converting them into enterprise-level solutions, creating frustration and diminishing returns.

Closely related is the **technology treadmill trap**, where organizations feel pressured to constantly chase the next AI breakthrough. In the race to adopt what's new, they often fail to extract full value from existing systems or integrate learnings into repeatable processes.

As AI collaboration expertise builds within the organization, another challenge emerges: **knowledge leakage**. Without systems in place to document, share, and preserve insights, companies risk losing valuable know-how as people move between teams or exit the organization entirely.

Perhaps the most subtle yet serious threat is the **cultural reversion risk**—the tendency for teams and leaders to revert to familiar patterns when stress levels increase. Under pressure, innovation often gives way to control, risk aversion, and rigid hierarchies, undermining the very conditions necessary to sustain AI collaboration.

To nurture a lasting impact, leaders must view these challenges not as barriers but as warning signals—evidence that reflection, reinforcement, and recalibration are essential to turning short-term momentum into long-term transformation.

Though these pitfalls may seem daunting, they are not insurmountable. With the strategies and supports outlined in the *Nurture* pillar of the **K.E.R.N. Human+AI Collaboration Framework®**, organizations can anticipate, address, and move past these challenges to create lasting value.

### Conclusion: Nurturing as the Keystone of Sustainable Human+AI Collaboration

As we draw this journey through the **K.E.R.N. Human+AI Collaboration Framework®** to a close, it's evident that *Nurture* is more than just the final pillar—it is the keystone that holds the structure of collaboration together. *Know*, *Empower*, and *Reflect* offer the necessary foundation and strength, but it is through *Nurture* that they are sustained and transformed into lasting impact.

Without ongoing nurturing, the insights from reflection diminish, the energy of empowerment fades, and even the strongest foundations weaken. True growth—whether in people, processes, or partnerships—demands intentional effort, patient investment, and leaders who are willing to foster resilience. Those who recognize this will view Human+AI Collaboration not as a one-time initiative but as a living ecosystem that must be cultivated to thrive.

Organizations that nurture both human and technological elements will be best positioned to achieve lasting innovation and competitive advantage.

Given the rapid changes and evolution of AI Technologies, as a thank you for your time and readership, I am providing all readers access to Transform or Die as a living, consistently updated reference resource where you will find the latest information about AI tools and platforms, practice prompts, easy-to-use scorecards, and self-directing training programs for your teams.

You will find a number of resources and checklists on the website to help you Nurture Human+AI on a long-term basis.

Use this QR code or visit RussellMKern.com/transform to access the resources.

# Chapter 12

# Transform or Die – The CEO's Guide to Human+AI Collaboration Success

*"Why can't our AI assistant just pull up our data and show us what's going on with the Wilson Project?"*

Jack, the CEO of ServicePlus, drummed his fingers on the conference table. His leadership team shifted uncomfortably in their seats. They'd invested heavily in an AI solution that was supposed to transform their operations. So far, it could write decent emails and answer basic questions, but when asked about actual company projects or data, it fell flat.

"We spent all this money, and it can't even access our own systems?" Jack continued, his frustration evident. "I thought AI was supposed to be revolutionary."

Sarah, the CTO, cleared her throat. "The problem isn't that AI can't do these things. It's that we implemented it without knowing our workflows, understanding our data, having our data ready, and then being able to build a solid business case. We bought a solution that can think but doesn't use our data and therefore can't help us transform strategically."

In that moment, he confronted the reality: **technology alone won't transform an enterprise. True change demands a new way of working together.**

## The Crossroads Facing CEOs

Today's CEOs are at a pivotal juncture. Agentic AI promises staggering opportunity—trillions in value, competitive reinvention, accelerated growth. Yet, despite unprecedented investment, many organizations remain stuck in "pilot purgatory," unable to deliver enterprise-level change. Meanwhile, nimble challengers harness AI and culture to upend industries with astonishing speed.

But for all the hype and hardware, one competitive advantage can't be purchased: authentic cultural transformation. Technology unlocks new possibilities, but it is habits, relationships, and shared values—the rewiring of leadership, teams, and communication—that fuel sustainable success.

The challenge for every CEO is no longer "if" AI will reshape your business, but how well you and your team are equipped to lead transformation and create an organization ready for what's next.

## Ten Big CEO Challenges—And a Human+AI Path Forward

It's about growth, agility, leadership pipeline, talent, capital, innovation, branding, customer centrality, sales and marketing alignment, and a data-driven culture. Today's top organizational challenges are more complex and fast-moving than ever before.

Strategic AI, deployed thoughtfully, doesn't replace the need for creativity, empathy, and judgment. It amplifies them, giving leaders and teams the intelligence, speed, and vision to anticipate change and act decisively. But true Human+AI Collaboration success requires more than new technology. It requires a new mindset.

## Why Collaboration, Not Just Technology, Is the Game-Changer

Legacy structures, rigid hierarchies, and outdated team models undermine the very cross-functional collaboration that AI accelerates. When team relationships are weak, even the best technology produces only incremental gains. It's no longer enough to automate workflows; forward-looking organizations must modernize how people connect, share, experiment, and co-create with AI as a trusted partner.

Unlocking the next era of success lies in reshaping collaboration, building trust, and linking the unique strengths of humans and technology.

## What AI Can and Cannot Do

AI is invaluable for processing information at scale, surfacing patterns, and automating tasks. However, human strengths—such as empathy, judgment, ethics, and vision—remain at the heart of innovative teams and purposeful leadership. As recent research confirms, the capabilities that matter most in the AI era are distinctly human and are becoming even more valuable as machine intelligence expands.

## The Mindset Shift: From Tool to Team Member

Servant leadership, emotional intelligence, and strategic vision are as vital as ever. However, the most effective CEOs now view AI not simply as a tech investment, but as a collaborative force—a knowledge partner, scenario analyst, and accelerator of both thinking and action.

Rather than confining AI to the back office or IT, modern leaders bring it directly into strategy, talent, branding, and sales—everywhere that speed, insight, and agility matter.

# The K.E.R.N. Human+AI Collaboration Framework®: Your Leadership Pathway

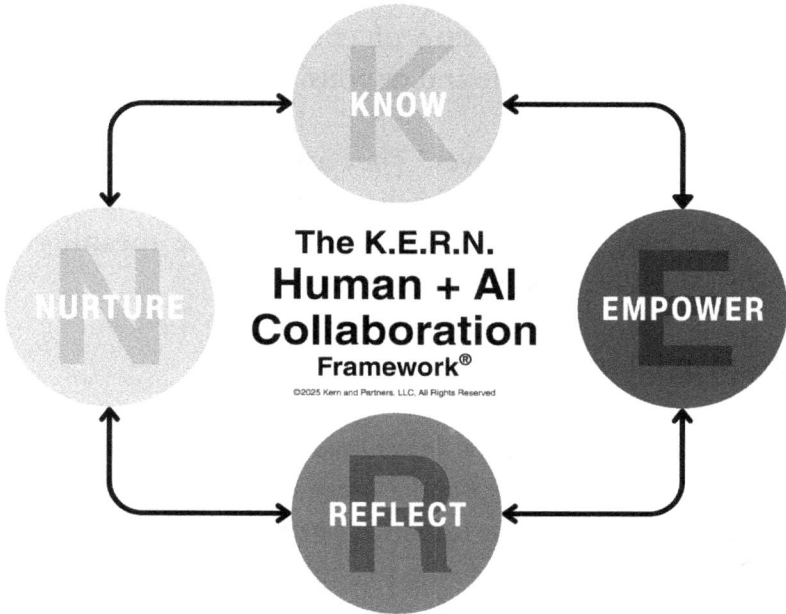

The K.E.R.N.
**Human + AI**
**Collaboration**
Framework®
©2025 Kern and Partners, LLC. All Rights Reserved

Earlier chapters explored the K.E.R.N. Human+AI Collaboration Framework® and presented it as a proven roadmap for cultivating high-performance Human+AI Collaboration:

**Know:** Build self and team awareness, identifying how human strengths and AI capabilities combine for smarter decisions.

**Empower:** Foster trust and psychological safety, supporting bold exploration, experimentation, and learning.

**Reflect:** Protect time for honest dialogue, candidly assessing team relationships, challenges, successes, and paths forward.

**Nurture:** Create a culture of continuous development, investing in both people and responsible AI governance.

These aren't "nice to haves." They are now mission-critical. When practiced consistently, these pillars enable your teams to move past pilot experiments and realize meaningful, unified transformation.

## Moving From Insight to Action

Transformation cannot be downloaded overnight. It requires vision, decisive leadership, and a commitment to the journey of growth, not as a one-time initiative, but as a continuous process. CEOs must set ambitious AI goals, connect fragmented teams, invest in future-ready skills, and model collaborative leadership. The blueprint for transforming your team is within these four pillars: lead with knowledge, empower with trust, reflect and adapt, and nurture for the long term.

## Your Living Resource for the AI Journey

As AI platforms, tools, and best practices continue to advance at a rapid pace, no single book can remain fully up-to-date. As a token of appreciation for your time and leadership, I am granting you exclusive access to *Transform or Die* as a living, always-current resource. Inside, you'll find the latest on:

- New and updated AI tools and platforms
- Practice prompts to build AI fluency
- Step-by-step, easy-to-use scorecards for identifying opportunities
- Guided self-directed training programs for your teams

Access this evolving toolkit by scanning the QR code or visiting RussellMKern.com/transform. Use it as your "forward operating base" for the next phase of Human+AI Collaboration. Return to it often for new insights, real-world case studies, and hands-on guidance.

## The Real Work Starts Now

This book is the starting line, not the finish. The organizations that thrive will be those that transform, with leaders who build environments that encourage creativity, connection, and daring reinvention.

The choice is yours: Lead transformation with purpose, or risk being left behind. The future belongs to those who act boldly today.

# Conclusion

# Continue the Journey Together

In a world marked by relentless change, enabling teams to grow and thrive demands more than just strategy—it requires sustained creativity, innovation, and a culture of high-value collaboration. This is precisely what the K.E.R.N. Human+AI Collaboration Framework® methodology is designed to cultivate.

If you are wrestling with a stream of frustrations:

*"Why am I the only one driving this?"*
*"Where are the fresh ideas?"*
*"Why isn't my team acting like a team?"*
*"Why don't my leaders know how to lead?"*
*"Why aren't my managers better coaches?"*

These are not signs of failure. They're signals that your team is hungry for alignment, clarity, and a new kind of leadership. And now, you're equipped with the roadmap to answer that call.

Imagine starting each day not with a jolt of caffeine, but with an inner certainty—knowing your team is aligned, your leaders are growing, and your organization is not just adapting to change but shaping it. That's what high-value Human+AI Collaboration unlocks.

You're building more than a high-performance business. You're nurturing a culture where teams bring their best, support one another, and focus relentlessly on delivering meaningful value together, every day.

## Human+AI Collaboration, as outlined in this book, will help you achieve this goal.

The Power of Transformational Human+AI Collaboration

- **Your team becomes a catalyst for growth.**
  When a team truly clicks, extraordinary things happen. They align with your vision, tackle challenges with creativity and resilience, and demonstrate problem-solving abilities that feel almost superhuman.

- **Your culture becomes magnetic.**
  A successful, energized workplace becomes a talent magnet. When people see teams engaged in meaningful work and empowered by continuous learning, they're drawn to be part of it, fueling your ability to attract and retain top-tier talent.

- **Your business value grows.**
  Transformational human-AI Collaboration yields measurable results, including expanded market share, increased sales, improved profit margins, stronger customer retention, enhanced competitive positioning, and sustained business growth.

- **Your teams prosper.**
  The true reward of transformational Human+AI Collaboration lies in the lived experience: deeper collaboration, mutual trust, amplified creativity, and a shared sense of ownership. Teams become agents of positive change, driven by inquiry, aligned in purpose, and proud of their progress.

Do you want to thrive—truly thrive—with a team that supports you, aligns with your objectives, and contributes meaningfully to your success? Isn't it time to cultivate a workplace where fresh ideas, innovative processes, breakthrough products, and creative solutions define your competitive edge?

**If you're ready to shift from frustration to fulfillment—and from stagnation to transformation—then let's talk.**

I invite you to connect with me directly to explore how we can apply the K.E.R.N. Human+AI Collaboration Framework® process within your organization. Together, we can build teams that not only perform but also drive real, measurable value. Let's elevate your culture into one that fuels growth, unlocks collaboration, and leads the future of Human+AI Collaboration.

Explore *Transform or Die* and use it as a living, continually updated reference hub. There, you'll find the latest tools, platforms, practice prompts, scorecards, and self-directed training programs to support your continued learning and team development in this new era of AI-powered collaboration.

Use this QR code or visit RussellMKern.com/transform to access the resources.

To contact me directly, please send an email to Russell@ kernandpartners.com or submit a contact request through my website at www.kernandpartners.com/info. If your matter is urgent, please call our offices at 877-323-5376.

# About the Author and His Firm.

**Russell M. Kern** brings over four decades of entrepreneurial leadership, creativity, and strategic innovation to the challenge of helping CEOs and their senior leaders transform how teams collaborate with one another and with AI.

Russell spends his time speaking, advising, mentoring, coaching, and providing professional, hands-on, live and virtual training for leaders and their teams on

- Why *Transform or Die* is a mandate within this 4th industrial revolution that has the potential to help businesses and their workforces thrive and flourish in unimaginable ways
- The risks and cost of avoiding the reset, rethinking, and mind shift to thrive in the AI age.
- The knowledge of the principles and application of processes within K.E.R.N. Human+AI Collaboration Framework® to transform current team dynamics into new productive norms that develop new levels of effective collaborative work between team members AND with AI as a strategic knowledge partner.

- Why the fundamentals of team leadership, including employee coaching, feedback, upskilling, and career growth mentoring, are and will always remain the separator between leaders and laggards, not technology.

As the founder and longtime President of a creative agency that grew from a two-person operation near LAX into a national powerhouse acquired by Omnicom in 2008, Russell guided 15 consecutive years of revenue growth while doubling the team from 200 to 400 members. Under his leadership, the agency delivered award-winning campaigns and breakthrough solutions for national brands including AT&T, American Express, Blue Cross/Blue Shield, Canon, DirecTV, SAP, and dozens more.

Russell's hands-on experience overseeing more than $1 billion in client marketing investments and orchestrating over 50,000 marketing tests shaped his deep understanding of what drives business results from cross-functional collaboration.

His commitment to understanding the science and psychology of high-performance teamwork has led him to advanced studies at leading institutions, including Neuroscience of Leadership at MIT, Creative Leadership at IDEO, and Appreciative Inquiry at Case Western Reserve University. Additional certifications include Working Genius, CliftonStrengths, and Learning & Development Operations through the Josh Bersin Academy.

Having experienced both the exhilarating highs of successful business exits and the humbling challenges of market disruption, Russell understands the real pressures facing today's business leaders.

**_Transform or Die_** is the culmination of his decades-long leadership in creative collaboration, now providing CEOs with the guidance to best enable new types of collaboration with AI technologies for the benefit of the company's stakeholders, investors, employees, partners, suppliers, and, most importantly, its customers.

Russell lives in California with his wife Meryl, his three dogs, Penny, Teddy, and Lol, and his two horses.

Russell's firm, Kern and Partners,' services include:

- Assessments of the current state of teamwork, collaboration, and AI fluency within teams, workgroups, divisions, or entire organizations.
- Reducing dysfunctional team dynamics in support of delivering new levels of customer satisfaction, business scale, and operational excellence.
- Custom design and implementation of organization-specific adult learning programming, proven to rapidly foster Human+AI collaboration behaviors.
- Helping leaders identify the highest and best use cases and design AI pilot programs that have the greatest likelihood of positively impacting the business in the long term.
- Facilitating data readiness, security guardrails, and technology access guidelines strategy sessions.

You can learn more about Russell Kern and his firm's approaches, as well as download tools, scorecards, and playbooks, and subscribe to his content.

If you have questions or want any other help, please contact Russell at Russell@kernandpartners.com or call his office at 877-323-5376.

Use this QR code or visit RussellMKern.com/transform to access these resources.

www.ingramcontent.com/pod-product-compliance
Lightning Source LLC
Chambersburg PA
CBHW071558210326
41597CB00019B/3297